W9-DEA-098

WITHDRAWN

PERGAMON INTERNATIONAL LIBRARY
of Science, Technology, Engineering and Social Studies
The 1000-volume original paperback library in aid of education,
industrial training and the enjoyment of leisure
Publisher: Robert Maxwell, M.C.

Sun Power

AN INTRODUCTION TO THE APPLICATIONS OF SOLAR ENERGY

THE PERGAMON TEXTBOOK
INSPECTION COPY SERVICE

An inspection copy of any book published in the Pergamon International
Library will gladly be sent to academic staff without obligation for their
consideration for course adoption or recommendation. Copies may be retained
for a period of 60 days from receipt and returned if not suitable. When a
particular title is adopted or recommended for adoption for class use and the
recommendation results in a sale of 12 or more copies, the inspection copy may
be retained with our compliments. The Publishers will be pleased to receive
suggestions for revised editions and new titles to be published in this important
International Library.

Other Pergamon Titles of Interest

ASHLEY et al.	Energy and the Environment—a Risk-Benefit Approach
BÖER	Sharing the Sun
BLAIR et al.	Aspects of Energy Conversion
BRATT	Have You Got the Energy?
DIAMANT	Total Energy
DUNN & REAY	Heat Pipes
HUNT	Fission, Fusion and the Energy Crisis
JONES	Energy and Housing
KARAM & MORGAN	Environmental Impact of Nuclear Power Plants
KARAM & MORGAN	Energy and the Environment: Cost-Benefit Analysis
KOVACH	Technology of Efficient Energy Utilization
MESSEL & BUTLER	Solar Energy
MURRAY	Nuclear Energy
SIMON	Energy Resources
SMITH	The Technology of Efficient Electricity Use
SPORN	Energy in an Age of Limited Availability and Delimited Applicability

Sun Power

AN INTRODUCTION TO THE APPLICATIONS OF SOLAR ENERGY

By

J. C. McVEIGH

M.A., M.Sc., Ph.D., C.Eng., F.I.Mech.E., M.I.H.V.E.

Head of the Department of Mechanical and Production Engineering, Brighton Polytechnic

PERGAMON PRESS

OXFORD · NEW YORK · TORONTO · SYDNEY · PARIS · FRANKFURT

U.K.	Pergamon Press Ltd., Headington Hill Hall, Oxford OX3 0BW, England
U.S.A.	Pergamon Press Inc., Maxwell House, Fairview Park, Elmsford, New York 10523, U.S.A.
CANADA	Pergamon of Canada Ltd., 75 The East Mall, Toronto, Ontario, Canada
AUSTRALIA	Pergamon Press (Aust.) Pty. Ltd., 19a Boundary Street, Rushcutters Bay, N.S.W. 2011, Australia
FRANCE	Pergamon Press SARL, 24 rue des Ecoles, 75240 Paris, Cedex 05, France
FEDERAL REPUBLIC OF GERMANY	Pergamon Press GmbH, 6242 Kronberg/Taunus, Pferdstrasse 1, Federal Republic of Germany

First edition 1977

Reprinted (with corrections) 1977

Library of Congress Cataloging in Publication Data

McVeigh, J C.
Sun Power.

(Pergamon international library of science, technology, engineering and social studies)
Bibliography: p.
1. Solar energy. I. Title.
TJ810.M2   1977   621.47   76-52963
ISBN   0-08-020863-0
ISBN   0-08-020862-2 flexicover

In order to make this volume available as economically and rapidly as possible the author's typescript has been reproduced in its original form. This method unfortunately has its typographical limitations but it is hoped that they in no way distract the reader.

Printed in Great Britain by William Clowes & Sons, Limited London, Beccles and Colchester

This book is dedicated to those who are striving
to conserve rather than consume, to develop a
simpler non-violent lifestyle and to obtain power
without pollution through the use of renewable
natural energy resources.

CONTENTS

PREFACE

We are living at a time when there is a greater awareness of the energy problems facing the world than at any other period in history. It is now widely accepted that the growth in energy consumption which has been experienced for many years cannot continue indefinitely as there is a limit to our reserves of fossil fuel. Solar energy is by far the most attractive alternative energy source for the future, as apart from its non-polluting qualities, the amount of energy which is available for conversion is several orders of magnitude greater than all present world requirements.

It is exciting and challenging to realise that we can all share in this inexhaustible energy source. In the long term some of the projects described in my book as being in their early stages of development will be providing a substantial part of the energy requirements of many countries. There are already several thousand solar water heating systems in the United Kingdom and each system is saving energy and helping the country to become more self sufficient in energy resources. The amount of money that a solar system saves annually will always keep pace with increasing fuel costs. By providing a general introduction to the very wide range of solar applications currently being pursued throughout the world I hope that many others will be encouraged to install their own solar systems, which will indirectly save many other resources such as water and building materials.

While a basic scientific knowledge would be an advantage for a complete understanding of each topic, the theoretical treatment has been reduced to a minimum and many of the applications are illustrated with original photographs, diagrams or sketches. The main emphasis is on thermal applications such as water heating, space heating and cooling and small scale power, which should make an immediate impact in many countries during the next five years. Practical construction details of several solar heaters and systems are included, as well as methods of assessing the economics of solar systems.

The applications are arranged in chapters which are intended to be complete in themselves and can be studied separately. This approach has meant that a few topics, such as the practical details of solar heating systems, appear in two chapters although the treatment is different. Architects and engineers will find that the reference sections, which contain over 350 references, are sufficiently comprehensive for immediate access to current research areas. Some of the references are to papers published originally as abstracts in the 1975 International Solar Energy Congress and subsequently in the journal Solar Energy in 1976. There is a considerable variation in the units used in the source material and generally these have been converted to SI units.

During the preparation of this book I have been helped either directly or indirectly by many people and it is not possible to mention them all individually. My original interests were in the thermal applications of solar energy and I am particularly grateful to those who have enabled me to have closer contact with different interests and new applications, especially my colleagues on the Committee of the UK Section of the International Solar Energy Society. Professor John Page, the first Chairman of the UK Section of

ix

ISES, has kindly allowed me to use some of his derived radiation data, and I
have benefited from frequent discussions with him on the wider issues.   It
was a rewarding experience to form part of his team which prepared the 1976
report on Solar Energy, a UK Assessment, and some of the material which
appears in Chapters 3, 4 and 5 is based on work which I prepared for that
publication.   I have also had helpful discussions with Professor David Hall
(Biology), Professor Peter Landsberg (Economics), Dr. Mary Archer (Photochemistry) and Michael Blee, Edward Curtis, Dominic Michaelis and Don Wilson
(Architecture and Solar Space Heating). Philip Baxendale suggested ideas for
my early experiments on domestic water heaters in the 1960's and technical
assistance has been provided by Roy Davis (Woolwich Polytechnic) and George
Moore and Dave Burton (Brighton Polytechnic).   Dr. Frances Heywood very
kindly allowed me to have access to the collected papers of the late
Professor Harold Heywood.

My thanks are also due to many friends and colleagues at the Brighton
Polytechnic.   Advice on collector materials and corrosion was given by John
Lane.   Dr. Alastair McCartney prepared Figure 2.2 and advised on solar
radiation data.   The eight full page illustrations were drawn by Leonore
Duff.   Garry Hibbert prepared many of the photographs.   Chris Wimlett traced
the line diagrams and Nancy Holmes typed a preliminary draft.   The final
typescript which appears in this edition was prepared by Celia Rhodes. I also
acknowledge the continued support for my solar energy work from the Director
and Council of the Brighton Polytechnic.

Some of the Meteorological data in Chapter 2 and the whole of Appendix 2
is reproduced with the permission of the Controller of Her Majesty's
Stationery Office.   I acknowledge with thanks the permission given by the
following organisations and individuals to use their illustrations:-   The
Copper Development Association (4.7), Edward Curtis and Homes and Gardens
(4.8), The Electricity Council Research Centre, Capenhurst (4.10), the
Philips group (4.15), Ferranti Ltd. (7.2) and Geoffrey Pontin and the Wind
Energy Supply Co. Ltd. (8.3).   Senior Platecoil Ltd. have allowed me to
quote from their brochure in Chapter 9.   The Editor of the Building Services
Engineer (formerly the JIHVE) has allowed me to quote directly from material
published in the June 1971 edition of the JIHVE.

Finally, it was my great privilege to serve as a Departmental Head under
the late Professor Harold Heywood at the Woolwich Polytechnic some ten years
ago.   He inspired many workers through the excellence of his research and his
helpful suggestions and enthusiasm laid the foundations for my own interests
in solar energy.

Brighton, Sussex   Cleland McVeigh
September, 1976

CHAPTER 1

INTRODUCTION AND HISTORICAL BACKGROUND

Man has appreciated for thousands of years that life and energy flow from
the sun. Socrates (470-399 B.C.) is believed to have been the earliest
philosopher to describe some of the fundamental principles governing the use
of solar energy in applications to buildings as the following passage from
Xenophon's Memorablia indicates:-

> In houses with a southerly aspect, the rays of the sun penetrate into
> the porticos during the winter, but in the summer the sun's path is
> directly over our heads and above the roofs, so that there is shade. If,
> therefore, this is the best arrangement, we should build the south side
> higher, to catch the winter sun, and the north side lower to exclude the
> cold winds . . .

Another early application was the alleged attack by Archimedes upon the
Roman fleet at Syracuse in the year 214 B.C. He is supposed to have constructed a number of highly polished focussing metal mirrors which were
arranged along the shore, so that the reflected rays of the sun were concentrated upon the hulls and rigging of the Roman ships which lay in the
harbour or sailed inshore. Some of them were set on fire and this caused the
Roman fleet to scatter. Most of the early applications related to various
focussing systems, such as mirrors or lenses. Anthemius de Tralles, a
celebrated architect in the sixth century, left among his papers four
treatises on burning mirrors. One of these treatises is entitled "How to
construct a machine capable of setting an object on fire at a distance by
means of solar rays". An English monk, Roger Bacon, also worked on burning
mirrors, towards the end of the thirteenth century. The first solar operated
water pump was invented by Salamon de Caus (1576-1626) a French engineer who
described this machine in 1615. The French philosopher Buffon carried out
various experiments in 1747 which demonstrated the practicability of the
attack at Syracuse. He build a large framework on which he hung pieces of
silvered glass which could reflect to a single focal point. He then varied
the number of mirrors and the focal point so that ultimately at a distance of
77 metres using 154 mirrors he burned chips of wood covered with charcoal and
sulphur. Subsequently he constructed a parabolic mirror with a diameter of
1.17 metres but all his experiments were regarded as, at best, scientific
curiosities by his contemporaries. An early example of a solar cooker is
recorded by De Saussure, a Swiss philosopher (1740-1799) in letters to Buffon
and La Journal de Paris. These describe how he had constructed a set of concentric glass chambers and cooked soup in the centre. A French physicist is
credited with a similar discovery at about the same time. Bernard Foret
Belidor (1697-1761) also invented a form of solar operated pump, or continual fountain, which is shown in Fig. 1.1.

Experiments to determine the intensity of the sun's radiation - the solar
constant - were first carried out at the beginning of the 19th Century by
Sir John Herschel, who invented an actinometer "an instrument for measuring
the intensity of heat in the sun's rays", and, quite independently, by the
French scientist Pouillet. Both used the same principle - exposing a known

1

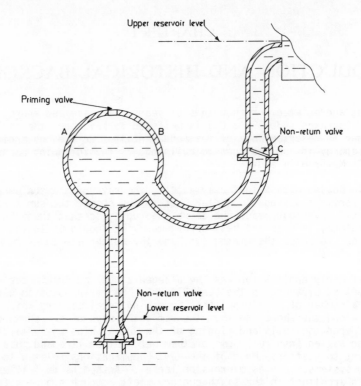

Upper reservoir level

Priming valve

A   B

Non-return valve

C

Non-return valve

Lower reservoir level

D

Fig. 1.1. Belidor's Solar Pump

*The pump is primed by filling the spherical dome to the
level AB. During the day solar radiation heats the dome,
causing the air to expand and force the water through the
non-return valve C to reach the upper reservoir. On
cooling, either artificially or at night, the internal
air pressure falls below atmospheric, drawing water into
the pump from the lower reservoir through the non-return
valve D.*

quantity of water to solar radiation and measuring the temperature rise which
occurred over a given period of time. Herschel's actinometer was a stationary open vessel while Pouillet's instrument was a closed movable vessel - the
pyrheliometer. Their calculations included allowances for atmospheric
absorption and scattering or attenuation. Pouillet's instrument and experimental methods were severely criticised by John Ericsson (1), who commented
that computations based on latitude, date and exact time were too complex and
tedious for investigations in which the principal element, the depth of the
atmosphere, was continually changing. Ericsson is better known for his
caloric engine and hot air cycle, but he was also a brilliant naval architect
before he turned to experiments with solar energy. His solar calorimeter was
fixed to "a vibrating table applied within a revolving observatory, supported
on horizontal journals and provided with a declination movement and a
graduated arc". On March 7th, 1871, he stated that "the dynamic energy
developed on one square foot of surface at the boundary of the atmosphere is
7.11 units (Btu) per minute". This is equivalent to 1.332 kW/m^2 - a figure

which is quite remarkable as it is within the allowable limits for the latest accepted estimates of the value of the solar constant.

The earliest recorded instance of a patent relating to solar energy (2) dates from 1854, when Antoine Poncon was granted a patent in London for ". . . using the sun's rays to create a vacuum in a suitable vessel, elevated at the height of a column of water, which, in the above vacuum, is kept in equilibrium by the pressure of the atmosphere. Such vacuum being formed, I fill it with water acted upon by the external pressure of the atmosphere, and thus obtain the heat of water which may be obtained as a motive power". Since then, nothing has been discovered about this invention and there are no records of its construction. Various other British patents occur over the next few years, but it is highly unlikely that the inventions which were claimed had ever been constructed. In contrast to these theoretical ideas, a French Professor, August Mouchot, had apparently constructed a parabolic focussing mirror which he used to drive a small steam engine in 1860 and he received a patent from the French Government in 1861. Subsequently he exhibited a "solar pumping-engine" in Paris in 1866 and also experimented with solar cookers. He wrote the first book ever published on solar energy, "La Chaleur Solaire et Ses Applications Industrielles" (3) in 1869, and on September 29th 1878 successfully demonstrated the first solar-operated refrigerator, producing a block of ice at the Paris Exposition.

Although Ericsson claimed in 1868 that he had constructed the first solar engines, it would seem that Mouchot was several years ahead of him. Ericsson was certainly the first inventor of a solar engine operating on an air cycle and this engine is reported to have worked in New York in 1872, "at a steady rate of 420 rev/min when the sun was in its zenith and the sky clear". Designed primarily for low powered applications, Ericcson noted that it was better to use steam in the engine when high powers were required.

With his considerable knowledge of solar radiation and his earlier experience in naval architecture and mechanical engineering, it is not surprising that Ericsson was also concerned about the energy crisis in 1876. He predicted that the coal fields would eventually be exhausted and that this would cause great changes in international relations in favour of countries with continuous sun power. He carried out a theoretical analysis on the use of solar engines on a strip of land about 8000 miles long and one mile wide (about 13000 km by 1.6 km) and commented:-

> Upper Egypt for instance, will, in the course of time, derive signal advantage and attain a high political position on account of her per-petual sunshine, and the consequent command of unlimited motive force. The time will come when Europe must stop her mills for the want of coal. Upper Egypt, then, with her never-ceasing sunpower, will invite the European manufacturer to remove his machinery and erect his mills on the firm ground along the sides of the alluvial plain of the Nile where sufficient power can be obtained to enable him to run more spindles than 100 such cities at Manchester.

Economic reasons also led to the development of the first, and, for many years, the World's largest solar distillation system in Las Salinas, about 110 km inland from the coast of Chile. The local water, which contained about 14% salts, was quite unsuitable for use in steam boilers and there was also the problem of supplying large quantities of drinking water for animals

and men. A complete specification of the system, which was designed by
Charles Wilson in 1872, was given by Harding (4). In consisted of 64 frames,
each 60.96 m long by 1.22 m broad, giving a total area of 4756 m^2 of glass.
A particular feature of the plant was an early application of self-
sufficiency, as the salty water was pumped from the local wells by a windmill
into a storage tank at the highest end of the plant. Initially some 19000
litres of fresh water could be produced daily at a cost of approximately one
quarter that of a conventional coal-fired boiler system, but after a railway
line had been built, the demand for the water decreased and the whole system
was allowed to disintegrate.

The earliest United States patent relating to some form of focussing
device was claimed by a parson, Charles Pope, in 1875. Pope was fascinated
by the wide variety of solar applications which were opening up at the time
and eventually produced and published in Boston the first book written in
English on solar energy in 1903 - "Solar Heat - its practical applications"
(2). The first major solar patent in the United States was issued on
March 20th 1877 to John S. Hittell and George W. Deitzler of San Francisco.
Their patent describes "a concave mirror by which they throw focalised heat
on a mass of iron or other suitable material as a reservoir of heat . . . .
letting the cold air pass in and then pass out again after the sun has heated
it, applying it then to ordinary hot air machinery" (the Ericsson cycle).
Deitzler took out a second patent on May 19th 1882 for a reflecting mirror
and was a founder director of the Solar Heat Power Company of California in
1883.

India was another country in which early work was carried out.
Mr. W. Adams, an English resident of Bombay, invented a solar cooker consisting of a conical reflector 0.711 m in diameter made of wood and lined
with common silvered cheap glass. "The rations of seven soldiers, consisting
of meat and vegetables, was thoroughly cooked by it in 2 hours in January,
the coolest month of the year in Bombay" (5).

Abel Pifre continued Mouchot's work in France and, on August 6th 1882,
used a mirror 3.5 m in diameter to power a small vertical steam engine which
worked a Marinoni printing press in Paris. Although there was some cloud in
the sky, an average of 500 copies per hour of a journal specially composed
for the occasion, "Soleil-Journal", were printed between 1 p.m. and 5 p.m.

One of the earliest space heating applications was claimed in 1882 by
Professor E.S. Morse of Salim, Mass. in an invention for "Utilizing the sun's
rays in warming houses" (6). It consisted of a surface of blackened slate
under glass, fixed to the sunny side of a house, with vents in the wall
arranged that the cold air in a room was let out at the bottom of the slate
and forced in again at the top by the ascending heated column between the
slate and the glass. This method was used to heat Professor Morse's own
house in fine weather. Also at about the same period the first use of the
flat plate collector is reported (7), but in an application to a water
pumping system.

The next thirty years, up to the outbreak of the first World War in 1914,
saw the size of the solar engine increase greatly. An anonymous group of
"Boston capitalists" developed several engines and, in 1901, their most
successful engine was described in various publications (8,9). The engine
was erected at an ostrich farm in South Pasadena, California and consisted of

a reflector, 10.2 m in diameter at the top and 4.57 m in diameter at the bottom, with an inner surface consisting of 1788 mirrors, approximately 90 mm x 600 mm, arranged to focus on a suspended boiler. The reflector stood on an equatorial mounting, with a north-south axis and tracked the sun, from east to west, by means of a clockwork mechanism. There is some doubt about its actual performance. Although an output of 15 hp has claimed, the actual recorded daily average when pumping water was only about 4 hp.

The other groups who were involved in large scale engine testing at that time were the Shuman Engine Syndicate Ltd., and the Sun Power Company (Eastern Hemisphere) Ltd. Their developments are very fully recorded by one of their consultants, A.S.E. Ackermann (10) in 1914. Mr. Frank Shuman's 1907 prototype consisted of a number of parallel horizontal black pipes containing ether, placed on the ground in a shallow box about 6 m x 18 m x 0.45 m deep, containing water with a layer of melted paraffin wax on it under a glass cover. The ether boiled and produced vapour at a sufficient pressure to drive a small vertical reciprocating engine. The exhausted ether was condensed and recycled. The second design was built at Tacony, Philadelphia in 1910 and was quite different in principle, using water only. A flat plate boiler was constructed from two thin copper sheets each 1.83 m long by 0.76 m wide with a narrow gap between them for the water. Cold water was admitted at the lower edge at one corner and a steam pipe was attached to the upper edge of the upper corner. The boiler was placed in a double glazed, insulated wooden box which was mounted on an east-west axis. No attempt was made to track the sun, but the inclination of the box was adjusted weekly so that the glass top was perpendicular to the sun's noon position. The system was successful in raising steam. The following year a full-size system was built, with 956.5 m^2 of collector area and the use of plain glass mirrors to achieve a concentration ratio of 2:1. No satisfactory method of measuring the actual output from this plant was available, but a maximum estimated 26.8 bhp was obtained by matching the steam conditions with the earlier test results.

The group then invited Professor C.V. Boys to join them in which became the most spectacular solar engine development of the time - the Shuman-Boys Sun-Heat Absorber at Meadi in Egypt. Professor Boys improved the Tacony design by introducing an automatic tracking system. The absorber consisted of five large parabolic mirror sections, each 62.5 m long and 4.1 m wide between the edges of the mirrors, giving a total collecting area of 1277 m^2. Each mirror was built up from various sizes of flat glass coated with shellac. They were carried on a light framework of painted steel and each of the sections was driven by a system of tubular shafting which caused the parabolic mirrors to rotate. They were placed with the major axes North-South. Each morning they were heeled over to the East and then they moved automatically and slowly from that position to the West, tracking the sun. An extensive series of tests were carried out in 1913, with a maximum recorded pumping horsepower of only 19.1 hp. Ackerman commented that this was an exceedingly bad result which he attributed to the engine and pumping side of the plant. Calculations which he based on the performance of another steam engine which he tested in England show that the steam conditions at Meadi could have developed 55.5 bhp.

At this point it can be seen that although only a restricted range of engineering materials was available, the basic principles of many practical applications of solar energy had been understood and that much of the work

had required very considerable technical expertise. However, the era of
cheap alternative fuel resources had commenced and the next two decades saw
a period of comparatively little interest in solar energy as the development
of first oil and then gas had priority. Fortunately some dedicated workers,
such as Dr. G.C. Abbot in the United States continued to carry out original
research, but it was not until the early 1940's that a resurgence of interest
in the utilisation of solar energy took place. This was encouraged by the
Godfrey L. Cabot bequest to the Massachusetts Institute of Technology to
promote solar energy research and from this renaissance the work spread to
research teams in many parts of the United States and to other countries of
the world. The first major symposium on wind and solar energy was held in
New Delhi in October 1954 (11) and the need to establish closer links between
the various countries led to the formation of the Association for Applied
Solar Energy, now the International Solar Energy Society (ISES). This has as
its purpose the encouragement of basic and applied solar energy research, the
fostering of the science and technology relating to applications of solar
energy and the compilation and dissemination of information relating to all
aspects of solar energy. The New Delhi Symposium was followed in November
1955 by two conferences in Arizona, the first, on basic research, was held at
the University of Arizona (12) and the second was a world symposium at
Phoenix (13), where a large variety of solar equipment was displayed,
including radiation measuring instruments, water and air heaters, cookers,
models of various solar houses, high temperature furnaces, water stills,
photovoltaic converters and several different types of engine, the largest
developing about 2.5 hp.

Several other conferences were held during the next 15 years. In 1961 the
United Nations held a symposium in Rome on new sources of energy (14) and
there was also an international seminar in Greece (15). The 1970 ISES
Conference in Melbourne was the last of the pre energy-crisis era.

Two major reports were published shortly before the UNESCO Conference,
"The sun in the service of mankind" in Paris, July 1973, the first from the
United States (16) and the second from Australia (17). Both reports highlighted the benefits which their respective countries could obtain from solar
energy applications. More recently, Ireland (18) and the U.K. (19) have
published their own solar reports. In July 1975 the largest conference on
solar energy ever held took place at the University of California, Los
Angeles (UCLA), with 265 technical papers presented, over 60 different
commercial exhibits and a total attendance estimated at more than 1700. The
conference was dominated by the United States, with 179 papers, although, if
the basis of comparison was the number of papers per capita, the 20 papers
from Australia were more significant.

Interest in solar energy research and development has continued to spread
very rapidly since then and an impression of the extent of world activity in
1976 can be obtained from Table 1.1, which has been largely compiled from
surveys published in the U.K. (20,21), U.S.A. (22) and the E.E.C. (23).
From the various sources quoted in these surveys it is clear that even this
table underestimates the position, as it proved difficult in some cases to
obtain a response to the requests for information. The classifications in
Table 1.1 are necessarily rather broad and may include economic or theoretical studies. The financial resources available from one country to another
sometimes differ by a factor of a thousand or more, but a feature common to
nearly every programme is an understanding that it does not take a substantial

TABLE 1.1

	1	2	3	4	5	6	7	8	9	10
ARGENTINA	X	X	X		X	X		X		X
AUSTRALIA	X	X	X		X	X		X	X	X
AUSTRIA	X	X		X				X	X	X
BELGIUM	X	X						X		X
CANADA	X	X	X	X		X	X	X	X	X
COSTA RICA	X	X	X	X			X			X
DENMARK	X	X	X				X	X		
ECUADOR	X	X								X
FINLAND								X		
FRANCE	X	X	X	X		X	X	X	X	X
GREECE	X	X		X	X					
INDIA	X	X	X		X	X	X	X	X	X
IRAN	X	X	X		X	X		X	X	X
IRAQ	X				X					X
IRELAND	X	X				X	X	X	X	X
ISRAEL	X	X	X				X			
ITALY	X	X	X	X	X			X	X	X
JAMAICA	X		X			X				
JAPAN	X	X	X	X			X	X		X
JORDAN	X	X	X		X	X				X
KOREA	X	X								
KUWAIT	X							X		
NETHERLANDS	X	X					X	X		X
NEW ZEALAND	X		X							X
NIGERIA			X							X
OAS*	X		X					X		X
PAKISTAN	X		X		X	X		X		
PAPUA NEW GUINEA			X							
SAUDI ARABIA	X	X			X					X
SOUTH AFRICA	X	X	X				X			X
SPAIN	X			X	X			X		X
SRI LANKA	X		X			X				X
SWEDEN	X	X					X			X
TURKEY	X	X	X		X	X		X	X	X
UK	X	X	X			X	X	X	X	X
USA	X	X	X	X	X	X	X	X	X	X
USSR	X	X	X	X	X	X	X	X	X	X
WEST GERMANY	X	X	X		X	X	X	X		X
WEST INDIES	X		X		X	X	X		X	X

Key to column headings
1. Water heating, including domestic and industrial processes
2. Space heating, solar houses, applications in architecture
3. Smaller thermal applications, pumping, engines, refrigeration and cooling
4. Large scale thermal applications, including focussing devices and furnaces
5. Desalination and distillation
6. Agricultural, including crop drying
7. Wind power
8. Photovoltaic and photochemical applications
9. Photobiology, bioconversion
10. Radiation studies

* OAS - The Organisation of American States

research and development programme to make a worthwhile contribution. The
world's largest programme started in the United States in 1971 with a modest
expenditure of 1.2 million dollars. Its growth to an estimated 300 million
dollars in 1977 is illustrated in Fig 1.2.

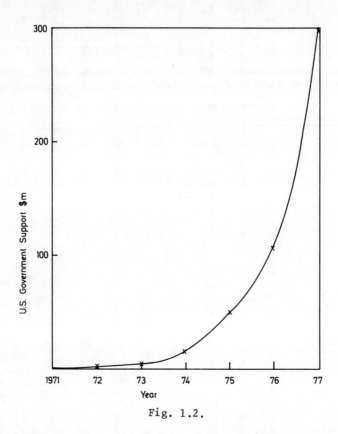

Fig. 1.2.

Conventional fossil fuel reserves can only last perhaps a hundred years at
the most and there are very considerable technical and environmental reserva-
tions about nuclear energy. Solar energy, which is non-polluting and
inexhaustible, is already economically viable for certain applications in
almost every country in the world. Political decisions to invest in solar
research, development and demonstration programmes have been taken in some
of the countries listed in Table 1.1. For those who are in a position to
influence national energy policies towards an increasing use of solar energy
there is only one message - the time is short.

References

(1) Ericsson, J., Contributions to the Centennial Exhibition, John Ross &
        Co., New York, 1876.

(2) Pope, C.H., Solar Heat - its practical applications, Boston,
        Massachusetts, 1903.

(3) Mouchot, August, <u>La Chaleur Solaire et Ses Applications Industrielles</u>,
     Gauthier-Villars, Paris, 1869.

(4) Harding, J., Apparatus for Solar Distillation, Paper No. 1933, Selected
     Papers, <u>Institution of Civil Engineers</u> 73 (1908).

(5) Adams, W., Cooking by Solar Heat, <u>Scientific American</u>, June 19th 1878.

(6) <u>Scientific American</u>, May 13th 1882.

(7) The utilization of solar heat for the elevation of water, <u>Scientific
     American</u>, October 3rd 1885.

(8) A solar motor, <u>The Engineering Times</u> 5, No. 4, 186-7 (April 1901).

(9) Thurston, R.H., Utilizing the sun's energy, <u>Cassiers Magazine</u>, New
     York (August 1901).

(10) Ackerman, A.S.E., The utilisation of solar energy, <u>Trans. Soc. of Engs.</u>
     81-165 (1914).

(11) Wind Power and Solar, Proc. New Delhi Symposium 1954, Paris, UNESCO,
     (1956).

(12) Trans. Conf. on the Use of Solar Energy - the Scientific Basis, Tucson,
     Arizona, November 1955, University of Arizona Press (1958).

(13) Proc. World Symposium on Applied Solar Energy, Phoenix, Arizona,
     November 1955, Stanford Research Institute, Menlo Park, California,
     (1956).

(14) U.N. Conf. on New Sources of Energy, Rome 1961. Proceedings, 4-6,
     New York, United Nations, (1964).

(15) Spanides, A.G. and Hatzikakidis, A.D., eds, Solar and Aeolian Energy,
     Proc. Int. Seminar, Sounion, Greece, September 1961, Plenum Press,
     New York, 1964.

(16) Solar Energy as a National Energy Resource, NSF/NASA Solar Energy Panel,
     December 1972.

(17) Report of Committee on Solar Energy Research in Australia, Australian
     Academy of Science, July 1973.

(18) Lalor, E., Solar Energy for Ireland, Report to the National Science
     Council, Dublin, February 1975.

(19) Solar Energy: a U.K. assessment, U.K. Section, I.S.E.S., London, May
     1976.

(20) McVeigh J.C., Advances in Solar Energy, <u>Heating and Ventilating News</u>,
     18 (9) (1975).

(21) Solar energy utilisation in U.S.A., France, Italy and Australia. Proc.
     Conf. Brighton Polytechnic, U.K. Section, I.S.E.S., July 1974.

(22) deWinter, F. and deWinter, J.W., eds, Description of the Solar Energy
      R & D programs in many nations, ERDA Division of Solar Energy,
      February 1976.

(23) Eggers-Lura, A., ed, Flat plate solar collectors and their application
      to dwellings, Commission of the European Communities study contract
      No. 207-75-9 ECI DK, Copenhagen, February 1976.

CHAPTER 2

SOLAR RADIATION

Radiation is emitted from the sun with an energy distribution fairly similar to that of a 'black body', or perfect radiator, at a temperature of 6000 K. Radiation travels with a velocity of 3×10^8 m/s taking approximately 8 minutes to reach the earth's atmosphere. The value of the solar constant - a term used to define the rate at which solar radiation is received outside the earth's atmosphere, at the earth's mean distance from the sun, by a unit surface perpendicular to the solar beam - is 1.353 kW/m². During the year the solar constant can vary by ± 3.4%, partly due to variations in the earth-sun distance.

The earth follows an elliptical path round the sun, taking about a year for each cycle. The earth's axis is tilted at a constant angle of 23° 27' relative to the plane of rotation at all times. The apparent daily motion of the sun across the sky viewed from any particular location on earth varies cyclically throughout the year and is defined by the angle of declination. This is the angle formed at solar noon between a vector parallel to the sun's rays which would pass through the centre of the earth and the projection of this vector upon the earth's equatorial plane. The angle of declination varies from + 23° 27' to - 23° 27'. This affects the angle of incidence of the solar radiation on the earth's surface and causes seasonal variations in the length of the day. At the equator the day lasts for exactly 12 hours from sunrise to sunset, but at higher latitudes there is a considerable variation. For example, in the British Isles the day lasts for less than 8 hours in mid-winter compared with 16 hours in mid-summer. This means during the mid-summer period the total radiation on a horizontal surface in the British Isles can be greater than in the equatorial regions.

Four dates in the year have a particular significance. These correspond to the two points in the earth's orbit when the effect of the earth's tilt is at a maximum, the solstices, and the two points when the tilt has apparently no effect, the equinoxes. In the northern hemisphere at the summer solstice, which occurs about June 22nd, the Arctic has continuous daylight as the north pole is at its closest position to the sun. Similarly in the southern hemisphere at the winter solstice, about December 22nd, the Antarctic experiences continuous daylight. The sun is directly over the Tropic of Cancer at noon at the summer solstice and directly over the Tropic of Capricorn at noon at the winter solstice. At the spring and autumn equinoxes, which occur about March 21st and September 23rd, the sun is directly over the equator at noon and, at any point on the earth's surface, day and night last for exactly 12 hours. Tables and charts referring to astronomical data normally relate to solar time, that is time relative to noon with the sun at the due south position in the northern hemisphere (or due north in the southern hemisphere). Solar time is often slightly different from local standard time as this can apply over several degrees of longitude and one degree of longitude is equivalent to four minutes of standard time.

11

Global, Direct and Diffuse Radiation

The availability of solar energy in any location in the world can be
studied by two methods. The first involves measurements from a radiation
monitoring network and the second is based on the use of physical formulae
and constants. Direct solar radiation, I, is the solar radiation flux
associated with the direct solar beam from the direction of the sun's disc,
which may be assumed to be a point source, and is measured normal to the beam
(that is on a plane which is perpendicular to the direction of the sun).
Diffuse radiation, D, reaches the ground from the rest of the whole sky hemisphere from which it has been scattered in passing through the atmosphere.
Global solar radiation, G, includes all the radiation, direct and diffuse,
incident on a horizontal plane. The distribution of diffuse radiation is not
uniform over the whole sky hemisphere and is more intense from a zone of
about 5 degrees radius surrounding the sun. This is known as circumsolar
radiation. Radiation may also be reflected from the ground onto any inclined
surface, though this is very difficult to assess. The relationship between
direct radiation, I, the diffuse radiation, D, and the global radiation, G,
is given by:-

$$G = D + I \sin \gamma \tag{2.1}$$

where γ is the solar altitude above the horizon.

Spectral Distribution of Direct Solar Radiation

The spectral distribution of direct solar radiation is altered as it
passes through the atmosphere by absorption and scattering. The amount of
energy absorbed depends on the length of path the solar beam traverses. A
common method of describing relative energy levels is the air mass, which is
the ratio of the actual length the solar beam traverses relative to the depth
of the atmosphere with the sun in its zenith position. Referring to Fig. 2.1,
the zenith path, ZO, is defined as unit air mass, the angle ZOS between the
zenith and the sun's direction is termed the zenith distance, z, and the air
mass, m, = SO/ZO = sec z, provided the curvature of the earth is neglected.
The second relation is very nearly equal to the true value allowing for
curvature up to 70^0 (air mass = 2.92). Beyond this it is necessary to allow
for variations in atmospheric refraction and to decrease of density with
height, as well as for curvature (1).

The spectral distribution curves for four different cases are shown in
Fig. 2.2. Two are for theoretical 'black body' radiation, the first at
6000 K and the second at 5630.7 K which is the sun's equivalent black body
temperature with the same overall radiation output as the solar constant (2).
The third curve shows the extra-terrestrial solar spectrum (2) and the fourth
and lowest curve represents the direct solar spectrum for a relatively clean
atmosphere, calculated for a zenith angle of 30^0 and equivalent to a typical
clear day in rural England at noon in summer (3). The aerosol (dust)
attenuation was calculated for a continental size distribution and the Ozone
and Rayleigh attenuation estimated from Elterman (4). Also from this lowest
curve it can be seen that the radiation is limited to wavelengths between the
near ultra-violet of 0.3 μm and the middle infra-red region of about 2.5 μm.
Absorption by gases and water vapour or cloud droplets occurs only in certain
specific narrow wavebands. The absorption of radiation by cloud is surprisingly small, perhaps less than 10% for a cloud 1000 m thick (5), but the main

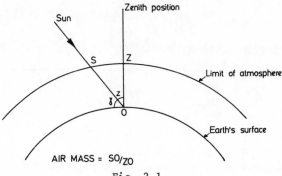

Fig. 2.1.

loss is due to scattering. Absorption by aerosol also occurs. Scattering of
radiation by cloud droplets and aerosols depends on wavelength and particle
size. With low particle concentrations the scattering tends to be forward,
giving relatively intense white diffuse radiation under hazy skies or thin
cloud. A very dense cloud 1000 m thick could reflect back into space more
than 90% of the incident solar radiation. The study of spectral distribution
is based on the use of physical formulae and constants and is of immense
importance in all photo-chemical and photo-biological applications.

The peak radiation received on earth is about 1.0 kW/m² normal to the beam
and the direct component of this is about 0.8 kW/m² when the sky is clear.
This gives an effective depletion to about 70% of the value of the Solar
Constant. It is very useful to bear this figure of 1.0 kW/m² in mind when
assessing the performance of a solar energy system, as any claims for an output from the system which approach the value of 1.0 kW/m² should be treated
with considerable caution.

Radiation Measuring Instruments

The earliest standard instruments for measuring direct beam radiation were
the Angstrom pyrheliometer developed in Stockholm and the Abbot water flow
calorimeter of the Smithsonian Institute in Washington. In the Angstrom
instrument the heating effect on a receiving element exposed to solar radiation is matched by an electrically heated shaded element. Standard methods
are used to measure the electrical heating. The Abbot water flow calorimeter
contains a cavity which absorbs solar radiation and the temperature rise in
the circulating cooling water is proportional to the solar irradiance. The
Abbot silver disc pyrheliometer is a secondary standard in which the rate of
change in the disc temperature is approximately proportional to the
irradiance. For many years it was noticed that the American and European
radiation measurements did not agree, with differences found by various
investigators in many countries, ranging from between 2.5% and 6% (6). In
September 1956 a new scale, designated the International Pyrheliometric Scale
1956, was approved which applied corrections of + 1.5% to the Angstrom scale
and - 2.0% to the Abbot, Smithsonian scale. Subsequently all instruments
since then have adopted the International Pyrheliometric Scale 1956.

Pyranometers, which are used for measuring global radiation, or, when
shaded from the direct beam, for diffuse radiation, are often based on the
principle of measuring the temperature difference between black (radiation

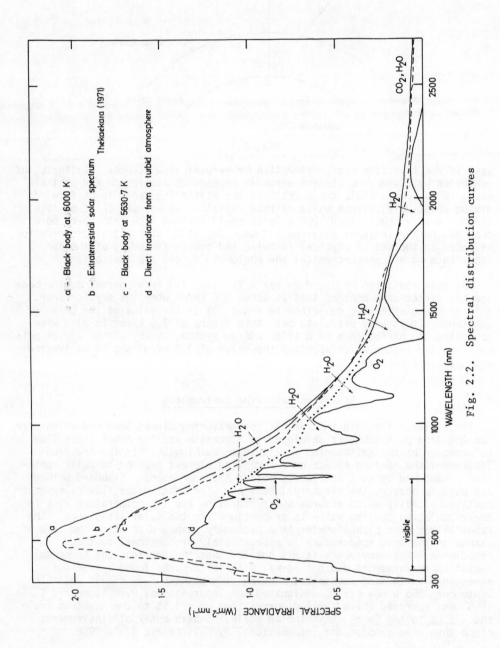

Fig. 2.2. Spectral distribution curves

a — Black body at 6000 K
b — Extraterrestrial solar spectrum
        Thekaekara (1971)
c — Black body at 5630·7 K
d — Direct irradiance from a turbid atmosphere

absorbing) surfaces and white (radiation reflecting) surfaces by the use of
thermopiles. These give millivolt outputs which can be conveniently handled
by a variety of conventional data recording systems and the Eppley pyranometer is a typical example of this type. Another well known type is the
Robitsch, based on differential expansion of a bimetallic element, while the
Bellani distillation pyranometer measures the global irradiance over a given
period of time by the distillation of alcohol with a calibrated condenser.
A much simpler type of measurement which is carried out in many locations is
the duration of 'bright sunshine'. This is measured by the Campbell-Stokes
recorder which uses a spherical lens to focus the solar radiation onto a
heat-sensitive paper which burns at the onset of 'bright sunshine'. This can
be correlated with the total global radiation by using the regression
equation:-

$$G = G_1 \left[a + \frac{bn}{N} \right]$$

where G = mean global radiation on a horizontal surface, G_1 = the reference
global radiation, n = mean value of the duration of bright sunshine, N =
mean length of day (or maximum possible daily value of bright sunshine) and
a and b are constants. The usual length of time considered for applications
of this formula is one month.

   A good example of the way in which the equation can be applied was given
by Connaughton's (7) analysis of radiation in Ireland where the data from
twenty-three stations recording bright sunshine was correlated with the
Valentia data from September 1954 to August 1965, giving values of a = 0.25
and b = 0.58. A series of maps giving the estimated mean global solar radiation for each month was then prepared. Similar work was carried out by
Day (8) for the whole of the British Isles. Day's work was more elaborate as
he found that the constants a and b varied widely from one station to
another. Variations can also occur from one period to another at the same
station, as Day's values for a and b at Valentia from 1954 to 1959 were 0.22
and 0.65 respectively.

Data from a Radiation Measurement Network

   It is difficult to obtain reliable solar radiation data. Even
experienced meteorological observers find accuracies of better than ± 5% in a
continuous series of long term observations hard to achieve. The most
reliable data is associated with the main meteorological stations*, but these
are often widely dispersed and a considerable distance from the location of
any potential application. Fortunately in the UK for most practical design
purposes it can be assumed that the averaged radiation data from any meteorological station within 150 km will be perfectly adequate. There is very
little systematic variation up to a distance of 300 km (9). This is
illustrated by reference to Table 2.1, where it can be seen that there are
two main trends - higher levels of global radiation towards the west, where
the skies are, in general clearer, and lower levels towards the north, which
would be expected because of the higher latitude.

   This data is for the period 1965-70, with the exception of Aldergrove
which is for 1969 and 1970 only, and was derived from Meteorological Office
records by Page (10).

---

*A complete list of the UK meteorological stations reporting hourly totals of
solar radiation is given in Appendix 2.

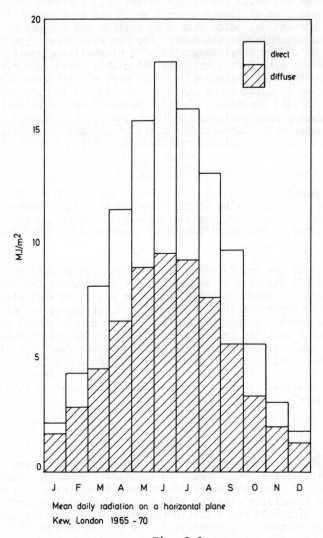

Mean daily radiation on a horizontal plane
Kew, London 1965 - 70

Fig. 2.3.

More than half the solar radiation received in the British Isles is
diffuse and this puts limitations on any applications which involve
focussing. Fig. 2.3 shows the six year average (1965-1970) of global solar
radiation at Kew, London, with the direct and diffuse components. The
comparatively low radiation levels in the winter period are combined with an
increased proportion of diffuse radiation, which greatly reduces the
effectiveness of many solar space and water-heating systems.

Applications in architectural design and housing often require a knowledge
of the total radiation on an inclined surface facing in any direction, while
the only available data is the total global radiation on a horizontal surface
in the same location or within a reasonable distance. Very few meteorological

TABLE 2.1

Annual variation of the mean daily totals of global solar radiation
on a horizontal plane (MJ/m²)

	Kew	Aberporth	Aldergrove	Eskdale Muir	Lerwick
January	2.13	2.39	1.67	1.54	0.82
February	4.13	5.03	4.50	4.36	2.97
March	8.06	9.51	7.29	7.30	6.41
April	11.62	14.25	12.22	11.47	12.21
May	15.54	16.89	14.65	12.99	13.60
June	18.06	20.09	19.48	16.57	16.90
July	16.03	18.13	15.35	13.64	15.42
August	13.29	15.08	13.45	12.24	11.93
September	9.73	10.58	8.86	7.79	6.88
October	5.79	6.16	4.39	4.52	3.54
November	3.00	2.97	2.66	2.41	1.37
December	1.72	1.94	1.46	1.37	0.55

stations give data for vertical irradiation, but this can be calculated and
Fig. 2.4, based on data derived for Dublin by Cash (11), shows the effects of
orientation on the ratio of vertical to horizontal irradiation.

The first work on the determination of the total radiation on an inclined
surface facing in any direction in the UK was carried out by Heywood (12, 13,
14) who suggested that calendar monthly radiation data should be replaced by
a system based on particular declination limits, having numerically equal
positive and negative magnitudes. The advantages of this system were claimed
to be as follows:-

   (i)   by dividing the year symmetrically about the summer
       solstice, periods of similar declination in the spring and
       autumn can be combined for the assessment of experiments or
       correlation with standard data;

   (ii)   the use of a relatively small number of standard declina-
       tion values reduces the amount of computation;

 (iii)   it would provide a better basis for the comparison of
       radiation data.

There was little support for this logical concept, but nevertheless Heywood
went on to establish parameters which can be determined from ratios of
radiation measurements, and produced curves showing how these could be
applied to determine the ratio of the incident radiation on any inclined
surface to the global radiation on a horizontal surface (15). These curves
were based on experimental measurements carried out on a continuous basis for
three years at the Woolwich Polytechnic (latitude 51° 30' N). Tables 2.2 and
2.3, based on these results, can be applied to the whole of the United
Kingdom provided that allowance is made at specific localities for variations
in the global radiation on the horizontal surface. The 'bright sun conditions' used by Heywood are those where the vicinity of the sun is free from

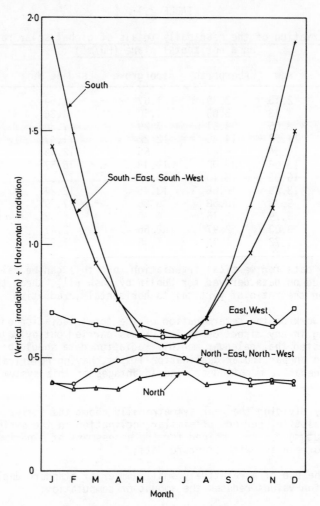

Fig. 2.4.   Effects of Orientation, Dublin derived
            data.

cloud and there is not more than one-third cloud cover.

Data from Kew in the period 1959-1968 was used by the Building Research
Establishment (16) to derive the monthly and annual totals of solar energy
incident on 1 m^2 of a surface inclined at different angles to the horizontal,
shown in Table 2.4.

It can be seen that over the whole year the total radiation prediction for
angles between 30^0 and 60^0 does not vary by more than a few per cent and that
in the summer months the smaller the angle the greater the total amount of
radiation. This theoretical analysis is confirmed when Table 2.3 is
examined.

An alternative approach to the problem of predicting the hourly solar
radiation incident upon any inclined surface was suggested by Bugler (17),
who used a mathematical model of solar radiation in which the diffuse
component is calculated from global horizontal radiation using three
different relationships, the appropriate equation being selected according to
the value of the ratio of measured hourly global insolation to hourly global
insolation computed for clear sky conditions. The method was checked with
reference to data for Melbourne for the period 1966-1970 with very good
results and is believed to have general application.

TABLE 2.2

Total radiation on South facing surfaces under bright sun conditions,
MJ/m^2 day

Position of surface	Winter period Oct 16-Feb 26	Spring & Autumn Feb 27-Apr 12 Aug 31-Oct 15	Summer Apr 13-Aug 30	Annual mean
Horizontal	5.60	15.84	28.73	17.03
S 20°	9.43	20.01	30.32	20.08
S 40°	12.21	23.13	29.28	21.49
S 60°	13.48	22.78	26.16	20.67
S Vertical	12.38	18.57	16.69	15.56

TABLE 2.3

Total radiation on South facing surfaces under average conditions,
MJ/m^2 day

Position of surface	Winter period Oct 16-Feb 26	Spring & Autumn Feb 27-Apr 12 Aug 31-Oct 15	Summer Apr 13-Aug 30	Annual mean
Horizontal	2.49	7.47	14.51	8.35
S 20°	3.28	8.52	14.96	9.09
S 40°	3.79	8.99	14.50	9.20
S 60°	3.81	8.52	12.51	8.32
S Vertical	3.52	6.47	8.57	6.19

TABLE 2.4

Monthly and annual totals of solar radiation on inclined surfaces,
MJ/m², (derived by BRE computer program from Kew average solar
radiation data for 1959-1968)

Month	Direct				Diffuse*			
	30°	45°	60°	90°	30°	45°	60°	90°
January	50	65	70	70	40	40	35	30
February	70	80	85	80	65	65	55	45
March	165	180	180	145	130	130	115	95
April	170	170	160	105	190	175	165	130
May	230	215	190	105	250	240	225	180
June	250	225	190	90	265	250	235	190
July	200	185	155	75	275	265	245	190
August	210	205	185	115	225	215	195	160
September	195	205	200	150	155	145	135	115
October	135	155	160	140	100	95	85	70
November	70	85	90	90	50	45	45	35
December	50	60	70	70	35	35	30	25
Annual	1795	1830	1735	1235	1780	1700	1565	1265

*Includes ground-reflected radiation

References

(1) Heywood, H., Solar energy for water and space heating, J.I. Inst. Fuel.
      27, 334-352 (1954).

(2) Thekaekara, M.P., Solar energy outside the Earth's atmosphere, Solar
      Energy 14, 109-127 (1973).

(3) McCartney, H.A., Private communication.

(4) Elterman, L., Atmospheric attenuation model, 1964, in the ultraviolet,
      visible and infrared regions for altitudes to 50 km, Air Force
      Cambridge Research Laboratories, Environmental Research Papers
      No. 46, AFCRL-64-740.

(5) Unsworth, M.W., Variations in the short wave radiation climate of the
      U.K., U.K. I.S.E.S. Conference on U.K. Meteorological Data and Solar
      Energy Applications, London, February 1975.

(6) Drummond, A.J. and Greer, H.W., Fundamental Pyrheliometry, The Sun at
      Work 3, No. 2 (June 1958).

(7) Connaughton, M.J., Global solar radiation, potential transevaporation
      and potential water deficit in Ireland, Technical Note No. 32,
      Department of Transport and Power-Meteorological Service, Dublin,
      1967.

(8)   Day, G.J., Distribution of total solar radiation on a horizontal
      surface over the British Isles and adjacent areas, The
      Meteorological Magazine 90, 269-284 (October 1961).

(9)   Monteith, J.L., Contribution to discussion on (5).

(10) Page, J.K., in Solar Energy. Memorandum by the U.K. Section,
      International Solar Energy Society. Select Committee on Science and
      Technology (Energy Resources Sub-Committee), Appendix 1., Part 1.,
      House of Commons Paper 156-i, HMSO, January 1975.

(11) Cash, J., Solar Energy and Buildings, Paper presented to Building
      Design Team, IIRS, Ireland, 5th December 1974.

(12) Heywood, H., Standard date periods with declination limits, Solar
      Energy 9, No. 4, (1965).

(13) Heywood, H., Solar radiation on inclined surfaces, Solar Energy 10,
      No. 1, (1966).

(14) Heywood, H., A general equation for calculating total radiation on
      inclined surfaces, Paper 3/21, International Solar Energy Conference,
      Melbourne, Australia (1970).

(15) Heywood, H., Operating experiences with solar water heating, JIHVE 39,
      63-69 (June 1971).

(16) Courtney, R.G., An appraisal of solar water heating in the U.K.,
      Building Research Establishment Current Paper CP 7/76 (1976).

(17) Bugler, J.W., The determination of hourly solar radiation incident upon
      an inclined plane from hourly measured global horizontal insolation,
      CSIRO, SESReport 75/4 (1975).

CHAPTER 3

WATER AND AIR HEATING APPLICATIONS

Solar energy can be easily converted into heat and could provide a significant proportion of the domestic hot water and space heating demand in many countries. One of the drawbacks in high latitude countries, such as the United Kingdom, is that there are many days in the winter months when the total radiation received will be too small to make any useful contribution. The most widely known and understood method for converting solar energy into heat is by the use of a flat plate collector for heating water, air or other fluids. The term 'flat plate' is slightly misleading and is used to describe a variety of different collectors which have combinations of flat, grooved and corrugated shapes as the absorbing surface, as well as various methods for transferring the absorbed solar radiation from the surface of the collector to the heated fluid. Many different types of collector have been built and tested by independent investigators over the past fifty years, the early work being carried out mainly in the United States (1,2), the United Kingdom (3), Australia (4), South Africa (5) and Israel (6). Tests were carried out in specific locations with wide variations in test procedures and in the availability of solar radiation. The main objective of these tests has been to convert as much solar radiation as possible into heat, at the highest attainable temperature, for the lowest possible investment in materials and labour (7).

The major British research in this field was carried out by the late Professor Harold Heywood, commencing with experimental work in 1947 on the characteristics of flat plate collectors (3). His earliest experiments were carried out on a small square collector, 0.093 m^2 in area. The heat collected was absorbed by means of water channels soldered onto the back surface of a blackened copper plate and the rate of heat absorption was determined for various numbers of glass plates and for different temperatures of collection. The somewhat simplified theoretical treatment which he established at that time is still used as the basis for some of the current design work on domestic flat plate collectors.

As well as these fundamental studies on the principles of heat collection Heywood built a solar collector of approximately 1 m^2 area which worked satisfactorily for many years in his home about 15 km S.-E. of London. The collector was constructed from two sheets of corrugated galvanised iron and was installed in a conventional thermosyphon system. The water capacity of the collector was 22.5l and the storage tank had a similar capacity, giving a total water capacity of about 45 l. for 1 m^2 of collector surface. This particular ratio of collector size to water capacity has featured prominently in subsequent experiments carried out in many different countries. Heywood's general conclusions, which are still very relevant, were as follows:-

(i)   Simplicity of construction must be an essential feature of water and space heating.

(ii) There is a considerable variation in the heat recovery rate from day to day in the UK.

23

(iii)   Satisfactory heat collection efficiencies are only obtained where
        there is prolonged and intense direct radiation. Cloudy periods cause
        a serious reduction in the collection efficiency and, while diffuse
        radiation can be partly effective, it has much less value than direct
        radiation.

(iv)   There is no direct correlation between "sunshine hours" as registered
        in many parts of the UK and heat recovery.

He also commended that the variation in solar radiation, which is rarely
identical for even two or three consecutive days, made experimental work
exasperating!

    In South Florida, USA, during the late 1930's, solar energy was the main
method used for providing hot water services to single-family residences,
blocks of flats and other small commercial buildings. A recent survey
carried out by Scott (8) 1974, showed that nearly all the systems relied on
the natural thermosyphon principle (see Chapter 9) and the collectors consisted of copper tubes soldered onto copper sheets, painted matt black and
enclosed in a galvanised steel casing. Evidence obtained from suppliers and
users confirmed that the collectors, considered alone, were very durable and
some had lasted completely trouble-free for over thirty years. Even freezing
conditions, encountered rarely in the Miami area, failed to damage collectors
made with soft copper. Users who had discontinued using their solar systems
did so for three main reasons:-

(i)   Damage caused by leaking main storage tanks.

(ii)   Insufficient hot water.

(iii) The considerable expense required to replace the storage tanks.

The problems relating to the tanks were caused by the combination of copper
tubing in the collectors and the steel storage tanks, leading to corrosion.
The increasing use of domestic hot water through the progressive introduction
of washing machines and dish washers meant that many systems could no longer
cope with the demand. This early experience has proved very valuable in subsequent collector and system design studies.

    Considerable practical experience was gained in Australia during the
1950's (4) and the Commonwealth Scientific and Industrial Research Organisation (CSIRO) subsequently published a guide to the principles of the design
and construction and installation of solar water heaters (9). At that time
they observed that the solar water heating industry had become established
in Australia and it was a practical and acceptable way of providing a
domestic water supply at a reasonable cost. Straightforward items of equipment developed by CSIRO and others could be bought as standard items from
suppliers throughout Australia. The first cost was greater than that of conventional installations but operating and maintenance costs were much lower.
The Australian research also showed that a simple solar installation could
provide adequate hot water for the needs of an average family throughout the
year, although it was more convenient, and in several places more economical,
to boost the solar heat with conventional heat sources. Many commercial companies started to manufacture and supply solar water heaters at that time but
very few survived - mainly because there was little demand for "off-the-shelf"

solar equipment. The few that managed to continue commercial operations did so by offering complete systems and by the mid-1970's were established as leaders in the field with new designs based on years of practical experience.

The Flat Plate Collector

The majority of flat plate collectors have five main components, as shown in Fig. 3.1. These are as follows:-

(i)   A transparent cover which may be one or more sheets of glass or a radiation-transmitting plastic film or sheet.

(ii) Tubes, fins, passages or channels integral with the collector absorber plate or connected to it, which carry the water, air or other fluid.

(iii) The absorber plate, normally metallic and with a black surface, although a wide variety of other materials can be used, particularly with air heaters.

(iv) Insulation, which should be provided at the back and sides to minimise the heat losses.

(v)   The casing or container which encloses the other components and protects them from the weather.

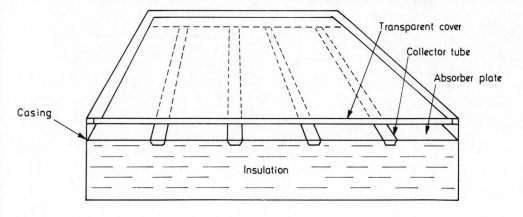

Fig. 3.1.

Components (i) and (iv) may be omitted for low temperature rise applications, such as the heating of swimming pools. Some of the very great variety of solar water and air heaters are illustrated in Fig. 3.2. Corrugated,

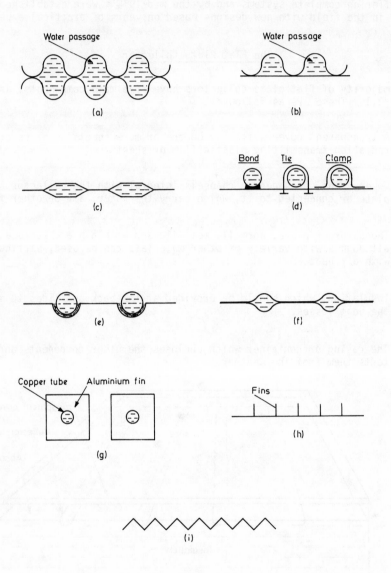

Fig. 3.2.   Cross-Sections through Collector
Plates

galvanised sheet is a material widely available throughout the world and
Figs. 3.2 (a) and (b) show two ways in which it has been used. The use of
conventional standard panel radiators (5,10), shown in Fig. 3.2 (c) is one of
the simplest practical applications (see Chapter 9). Methods of bonding and
clamping tubes to flat or corrugated sheet are shown in Figs. 3.2 (d) and (e)
while Fig. 3.2 (f) is the "tube-in-strip" or roll bond design, in which the
tubes are formed in the sheet, ensuring a good thermal bond between the sheet
and the tube. An effective commercially available collector is shown in
Fig. 3.2 (g), based on standard refrigeration heat exchanger practice. Two
different types of solar air heater surface are shown in Figs. 3.2 (h) and
(i).

   Flat plate collectors could also be classified into three groups according
to their main applications as follows:-

   (i)   Applications with a very small rise in temperature, such as in
         swimming pools where the collector needs no cover or insulation at the
         back or sides (11). A high rate of flow is maintained to limit the
         temperature rise to less than 2°C.

   (ii) Domestic heating and other applications where the maximum temperature
         required is no more than 60°C. Insulation at the back and at least
         one transparent cover are necessary.

   (iii) Applications such as process heating or the provision of small scale
         power, which temperatures considerably above 60°C are necessary. A
         more sophisticated design approach is needed to reduce heat losses
         from the collector to the surroundings.

   From the great variety of successful collectors shown in Fig. 3.2, the
flat plate collector appears to be a comparatively straightforward piece of
equipment and in an ideal collector all the radiation reaching it would be
converted into heat. In practice, the useful heat collected, Q, is always
less than the incident solar radiation, G_c. There are many different
factors which can contribute to this and a detailed analysis of the thermal
characteristics of the flat plate collector is very complex. For example,
the loss of heat by radiation increases as the fourth power of the absolute
temperature, making losses due to radiation increasingly significant as the
temperature of the heated fluid becomes more than 25°C above the temperature
of the surroundings. The first detailed analysis of these various factors
was carried out by Hottel and Woertz in 1942 (2). However, a comparatively
simple analysis will give very useful results and show how the important
variables are related and their effect on collector performance.

The Hottel-Whillier-Bliss Equation

   The basic equation, widely known as the Hottel-Whillier-Bliss equation
(12-17), expresses the useful heat collected, Q, per unit area, in terms of
two operating variables, the incident solar radiation normal to the collector
plate, G_c, and the temperature difference between the mean temperature of the
heat removal fluid in the collector, T_m, and the surrounding air temperature,
T_a, as follows:-

$$Q = F\{(\tau\alpha)G_c - U(T_m - T_a)\} \qquad (3.1)$$

where F is a factor related to the effectiveness of heat transfer from the collector plate to the heat removal fluid. This factor is influenced by the design of the collector plate, for example the dimensions of the passages containing the heated fluid and the thickness of the plate, as well as the properties of the fluid. It also varies with rate of fluid flow through the collector.

The transmittance-absorptance product $(\tau\alpha)$ takes account of the complex interaction of optical properties in the solar radiation wavelengths (17). It is actually some 5% greater than the direct product of the transmittance through the covers, τ, and the collector plate absorptance, α, because some of the radiation originally reflected from the collector plate is reflected back again from the cover.

The heat loss coefficient, U, rises very rapidly with increasing wind velocity if there are no covers, but is less dependent when the collector has at least one cover. The number and spacing of the covers and the conditions within the spaces can be significant, for example an evacuated space greatly reduces heat losses. The longwave radiative properties of the collector plate and covers also influence the heat loss coefficient.

These three design factors, F, $(\tau\alpha)$ and U, define the thermal performance of the collector and the overall efficiency of the collector, $\eta = Q/G_c$, can be expressed in terms of the temperature difference, $(T_m - T_a)$, and the incident solar radiation, G_c, in equation 3.2:-

$$\eta = \frac{Q}{G_c} = F(\tau\alpha) - \frac{U}{G_c}(T_m - T_a) \qquad (3.2)$$

The temperature T_m is almost impossible to measure, but as most systems have comparatively small temperature rises through any individual collector, the inlet fluid temperature T_i can usually be substituted for T_m. Typical values for F are about 0.88 to 0.90, $(\tau\alpha)$ for two covers of 3 mm window glass and an α value of 0.90 is about 0.7, while U for the same collector would be about 3.6. An unglazed, uninsulated collector would have a $(\tau\alpha)$ value approaching unity, but a U value at least twice that of the glazed collector. Experimental methods for determining F, $(\tau\alpha)$ and U are given by Smith and Weiss (17).

Physical Design Characteristics

The three design factors are all affected by the actual physical design characteristics of the collector, the main characteristics being the type and number of transparent covers and the properties of the collector surface. The wavelength range of the incoming solar spectrum is less than 3 μm for all than about 2% of the total incoming solar energy outside the earth's atmosphere. When this radiation reaches a sheet of glass, as much as 90% can be transmitted directly, the remainder being reflected or absorbed by the glass. The absorbed energy raises the temperature of the glass so that, in turn, the glass re-radiates from both the internal and external surfaces. As the surface temperature of the collector plate rises, it also radiates, but at greater wavelengths than 3 μm for all but a very small proportion of the total energy, typically less than 1% for a black surface at 100°C. The longwave radiation emitted by the collector plate cannot pass back directly through the glass, as the transmittance of glass is practically zero in the range 3 - 50 µm. This phenomenon is the well-known 'greenhouse effect' and the use of a transparent cover or several covers greatly reduces heat losses from the collector. Transparent plastic materials also possess high short-wave transmittance characteristics, but generally have appreciable longwave transmittances. With direct radiation the transmittance varies with the angle of incidence as shown in Fig. 3.3, where transmittance values for single and double glazing using double-strength clear window glass (18,19) are compared with a fibre glass plastic material (20). This particular material has exceptionally good properties in the longwave region, as shown in Fig. 3.4.

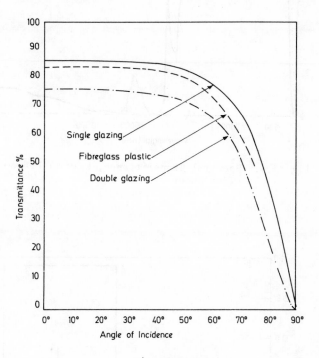

Fig. 3.3.

Each transparent cover reduces the outward heat losses from the front of the collector, but also reduces the total amount of incoming solar radiation which can reach the collector plate surface. When the absorption of energy into each cover is taken into account, the transmittance losses for incidence angles up to 35° are approximately 10% for single glazing, 18% for double glazing and 25% for triple glazing (19). The use of a combination of an outer glass cover with an inner, cheaper transparent plastic film can have advantages, as the plastic can have a higher transmittance than glass and the outer glass cover provides a certain amount of protection from weathering (21). The distance between covers or between the inner cover and the collector plate surface is not very critical (16). An optimum spacing of between 10 to 13 mm has been suggested (22), but up to 25 mm can be used.

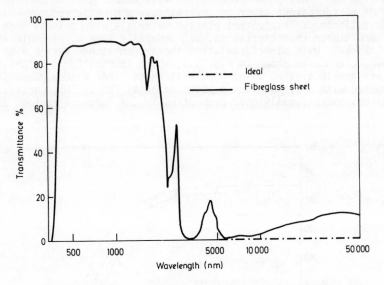

Fig. 3.4. Spectral Transmittance, Kalwall Fibre-
glass Sheet

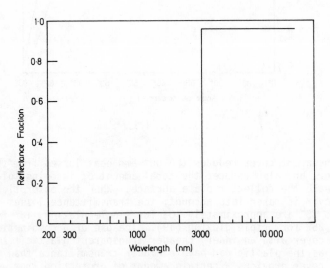

Fig. 3.5. Spectral Reflectance of Ideal Surface

The performance of a glass cover can be improved by depositing a transparent coating on its inner surface which allows nearly all of the incident solar radiation to be transmitted, but reflects any longwave radiation back to the emitting surface of the collector plate. Indium oxide and tin oxide are commonly used and a Japanese vacuum formed coating (23) gave a transmittance of 0.85 in the visible range (0.55 µm) and a reflectance of about 0.97 in the infra-red (4.0 µm). The early figures quoted for the Philips evacuated tubular collector (24) gave values of 0.85 and 0.9 respectively for τ and α.

Selective Surfaces

The longwave radiation emitted from the collector plate surface can be considerably reduced by treating the collector surface to reduce its emissivity in the longwave spectrum without greatly reducing the absorptivity for shortwave radiation. This concept is shown in Fig. 3.5, where the properties of an ideal selective surface are illustrated. The monochromatic reflectance is very low below wavelengths of 3 µm, known as the cut-off or critical wavelength, and very high above this value. For the great majority of flat plate collectors the temperature of the surface will be sufficiently low for practically all the emitted energy to occur in wavelengths greater than 3 µm. The contrast between the ideal surface and some real surfaces is shown when Fig. 3.6, adapted from McDonald (25), is examined. Real selective surfaces do not show the sharp rise in reflectance at one particular cut-off wavelength and their properties vary with wavelength. Complete integration over the emitted spectrum is necessary to estimate the long wavelength emittance and over the solar spectrum to estimate the solar absorptance.

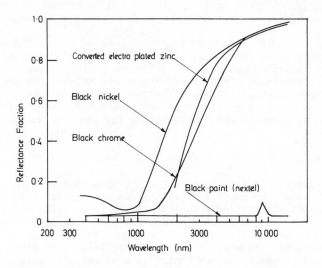

Fig. 3.6. Spectral Reflectance of Plated zinc

The effect on the performance of the collector of increasing the number of covers and applying a selective surface is illustrated in Table 3.1, which gives the overall loss to the surroundings from the top cover, assuming a mean wind velocity of 5 m/s and an ambient temperature of 10°C. The collector plate is considered at 40°C, a typical summer temperature achieved in the British Isles, and at 80°C, a temperature suitable for many process heat applications. Back and side losses from the collector casing are not considered. The figures are based on data from Duffie and Beckman (16).

TABLE 3.1

Overall Top Cover Loss (W/m^2)

Plate temperature	40°C		80°C	
Longwave emissivity	0.95	0.1	0.95	0.1
1 cover	189	93	525	263
2 covers	78	57	280	168
3 covers	63	45	182	119

The reduction in energy loss due to the selective surface becomes increasingly significant as the temperature of the collector plate rises. Any reduction in energy loss improves the collector efficiency and the overall annual increase in collected energy which can be achieved by the use of selective coatings depends on the number of hours per year during which there is sufficiently intense radiation to allow the collector to reach temperatures at which this becomes significant. An increase of up to 20% in the amount of energy collected in a year in the United Kingdom has been suggested (26).

The use of a second cover has almost the same effect on the top cover loss as a good selective surface in this temperature range, but this also reduces the amount of radiation reaching the collector plate surface. At comparatively small temperature differences between the collector plate surface and the surroundings, a single glazed collector is normally more efficient for this reason. The use of a selective surface and two covers give a comparatively small improvement over the selective surface with one cover.

Surface Treatment

There are several different ways in which selective surfaces can be prepared, depending on the physical principle involved. Tabor (27) has been concerned with finely divided metals deposited on polished metal undersurfaces; other methods include the deposition of thin semiconductor layers which absorb the short wavelengths but not the long, thus allowing the metallic undersurface to maintain its low emissivity. Surfaces may also be given a controlled degree of roughening so that only the absorptivity for short wavelengths is increased. Surfaces with large (relative to all radiation wavelengths) Vee-grooves can be arranged so that radiation from near normal directions to the overall surface will be reflected several times in the grooves. An effective α/e of 9 with $\alpha = 0.9$ has been suggested (16).

A method for the commercial production of a selective surface on a copper surface was described by Close (30). It consisted of dipping the copper plate in a solution of 1 part sodium chlorate ($NaClO_2$) to 2 parts sodium hydroxide (NaOH) in 20 parts of water by weight. The plate should be immersed for ten minutes at a controlled temperature of about 62°C. The usual precautions of ensuring that the plate was clean and degreased prior to immersion were recommended. The performance of a single glazed collector treated with this surface was found to be about 10% better in overall collection efficiency compared with a conventional non-selective double glazed collector.

Some chemically applied coatings referred to in the literature include a nickel-zinc-sulphide complex known as 'nickel black' (27), copper oxide on copper (27,28) and copper oxide on aluminium (29). A black chrome coating, the Harshaw ChromOnyx* process, was considered to be one of the best available commercial systems in 1975 (25,31,32). This is a modified version of a well-known standard decorative black chrome electroplate seen, for example, on office furniture. A comparison between some black chrome processes and other solar selective coatings is given in Table 3.2.

TABLE 3.2

Coating (and reference)	Absorptance (α)	Emissivity (e)	Ratio of absorptance to emissivity (α/e)
Nickel Black on galvanised iron (experimental) (27)	0.89	0.12	7.42
Same process (27)	0.89	0.16-0.18	5.56-4.94
Sodium hydroxide, Sodium chlorate on copper (30)	0.87	0.13	6.69
Black Chrome on dull nickel (31)[+]	0.923	0.085	10.86
Black Chrome on bright nickel (31)[+]	0.868	0.088	9.86
Black nickel (31)[+]	0.867-0.877	0.066-0.109	7.95-13.29
Nextel black paint (31)[+]	0.967	0.967	1.00

[+]Results based on spectrum weightings for solar air mass 2 (absorptance) and 121°C black body (emissivity).

*Registered Trademark.

A selective surface with a high ratio of absorptance to emissivity
(α/e = 20), with α nearly 1, has been obtained with a gold black coating
(33,34) placed on a reflecting undersurface such as copper. A relatively
inexpensive electrochemical method which uses a chromium based oxide coating
known commercially as 'Solarox' has been developed in Australia (35).
Typical results gave an α/e value of 18 at 25°C falling to 7.5 at 300°C.

Collector Materials and Corrosion

All components and materials used in a solar energy collector should be
designed to operate satisfactorily under the worst conditions which could be
expected in any particular installation. Materials should be capable of
withstanding both the high temperatures which would be encountered during
periods of maximum radiation with no flow through the collector and the low
temperatures which could occur in mid-winter. Problems which could arise
from cyclic variations in temperature or large temperature differences within
the collector should be taken into consideration in materials selection and
design. The estimated life of any component is important in determining the
real cost of the delivered energy and corrosion may be the greatest limiting
factor.

The majority of solar heating systems have more than one metal in contact
with the heat transfer fluid, which is usually water. Pipework may be copper
or stainless steel and the collector plate could be constructed from copper,
stainless steel, mild steel or aluminium. The presence of mixed metals in a
system is one of the most important mechanisms which can give rise to
corrosion. The other is the presence of dissolved oxygen in the heat
transfer fluid (36). In the simplest solar water heating system, the water
from the cold tank passes through a solar collector to preheat the feed to
the hot system. Oxygen is freely available in such a system and the wrong
choice of materials for the pipework and collector could result in a very
short life before perforation occurs. For example, an aluminium collector
panel tested with a steady flow of ordinary mains water containing dissolved
copper failed in less than two months (37). There are also the further
problems of scale formation in hard water areas and possible frost damage in
winter.

In closed circuit systems a heat exchanger is placed in the solar hot
water storage tank and the water recirculates through the collector and heat
exchanger. Standard finned copper tube can be used as the heat exchanger and
an experimental unit tried at Brighton Polytechnic is shown in Fig. 3.7.
These systems usually rely on initial corrosion to reduce the oxygen content
in the water to an acceptable level. A pump is often incorporated in the
circuit and the wrong positioning of the pump could send water up through the
expansion tank, picking up fresh oxygen. Ideally the expansion tank should
be sealed. Bacterial activity can also cause problems as some antifreeze
solutions when heated provide ideal conditions for fungal growth, especially
if the mains water used to charge the system contains dissolved salts.
Biocides can reduce this activity and the dissolved salts could be eliminated
by using de-ionised or distilled water. Mixed metal systems allow the
maximum use to be made of less expensive materials such as uncoated mild
steel, but they can only be used safely in a closed circuit with the addition
of a suitable inhibitor to prevent cuprosolvency and a biocide. If glycol is
added for frost protection it should be a suitably inhibited grade. The use
of a sealed closed circuit may be more acceptable to some local water

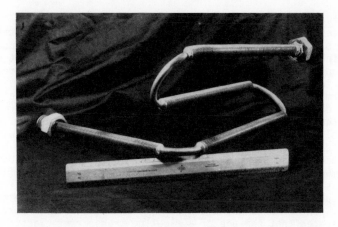

Fig. 3.7.

authorities who would object to the use of inhibitors with systems directly
connected to a mains water supply.

The estimated useful life of a selective surface is hard to assess and the
initial values of both α and e may degrade with use. Measurements which have
been made on some selective surfaces have shown that e increases with long
term use (38). Two possible causes are suggested for this, ultra-violet
radiation or the effects of atmospheric moisture and pollution. The insulation in any collector should have a low thermal conductivity and be thermally
stable at the maximum collector temperature. The various materials assembled
in a collector affect its thermal performance, but these effects are not
independent and must be evaluated in each specific application. The temperature level at which the collectors would be operating is particularly
significant, for example, a selective surface could be shown to be cost
effective in one case but not in another.

A comprehensive review of the problems associated with the use of
aluminium and copper was given by Popplewell (39). It includes a discussion
of system design to avoid corrosion and presents corrosion data in fresh
water for various copper alloys. The use of an organic, non-corrosive fluid
as the heat transfer medium was considered to be an acceptable alternative.

Developments in Collector Design

There are many different approaches to solar collector design and it is
only possible to examine some of the more interesting trends in detail.
Improvements in the overall efficiency of collection, particularly with
higher temperature differences between the heated fluid and the surroundings,
can seldom be achieved without increasing the complexity and cost of the
collector. For low temperature rise applications the emphasis is on designs
which can be shown to have very short payback periods (the capital cost of
the system divided by the current annual value of the fuel saved) in the
order of five years or less.

Low Temperature Rise Applications

At present the main application is in swimming pool heating, but there are
many other potential applications for this type of collector, such as in the
glasshouse industry and in fish-farming. The cheapest, simplest and most
direct method of heating any outdoor swimming pool is by the direct absorption of the incident solar radiation on the surface of the pool. With no
form of swimming pool cover or any other means of preventing heat loss from
the surface of the water the temperature of the pool in a temperate climate
such as in the British Isles would normally follow the mean air temperature
fairly closely during the summer months. However, the summer swimming season
can be extended by one or two months at either end by providing a solar water
heating system in addition to the direct absorption mentioned above. Another
extremely important factor is the reduction of the heat losses from the pool.
The heat loss by evaporation is the most significant (40), but fortunately
this can be almost completely eliminated by the use of a single thin cover on
the surface. Experiments in Australia (41) and at the University of Florida
(42,43) indicated that the use of a floating transparent plastic cover could
raise the average pool temperature by over $5^{\circ}C$ compared with a similar
identical unheated pool. A smaller temperature rise can be obtained in the
British Isles, partly due to the poorer radiation climate and partly to the
greater rainfall causing the cover to be partially submerged and reducing its
effectiveness. The other main losses are through convection and radiation.
Pools placed with exposed walls above ground level are usually colder than
conventional pools. Conduction losses from conventional pools can be
neglected as practically all the heat that escapes into the ground returns
again to the pool when the pool temperature falls (42).

For these low temperature rise applications, where the temperature rise is
only 1 or $2^{\circ}C$ to reduce heat losses from the collector, a simple unglazed
uninsulated collector is quite adequate and many designs have been based on
the use of black corrugated sheets with water flowing down the channels from
a perforated pipe. Known as the 'trickle' type, they have been used
extensively in the United States by Thomason, and some applications are
described in Chapter 4. Various types of black sheeting could be placed over
the corrugations or a single galvanised sheet can be painted black and
wrapped in transparent plastic, with the water trickling down both the front
and the back of the sheet (44). For high efficiency a uniform thin film of
water is desirable and a method for achieving this is described in Chapter 9,
where the black sheet is placed over a sheet of polythene packing material
containing a uniform matrix of equally spaced cylindrical air bubbles. The
water flows between the two sheets. If higher outlet temperatures are
required a transparent cover is needed and Fig. 3.8 shows an experimental
collector covered with a transparent plastic cover. Several commercial
manufacturers have now adopted these covers as their standard material for
both low and medium temperature rise applications.

Flat Plate Collectors

The thermal trap collector. This system was first proposed by Cobble (45)
and has been developed by the New Mexico State University (46). It uses a
transparent solid (methyl methacrylate) adjacent to the conventional flat
collector plate, as shown in Fig. 3.9. Methyl methacrylate has a high transmittance in the visible and shorter infra-red spectrum combined with a very
low transmittance at longer wavelengths and a small thermal conductivity.
Comparative tests carried out at New Mexico State University showed that the

Fig. 3.8.

thermal trap collector had superior characteristics to a conventional flat
plate collector and a trickle type collector. All three collectors were
tested at operating temperatures between 38°C and 80°C and in this range the
thermal trap collector had a higher collection efficiency and could operate
usefully for more hours in the day. It was less affected by intermittent
cloud conditions as it had a relatively large time constant and it appears to
be very promising for use as a high temperature collector.

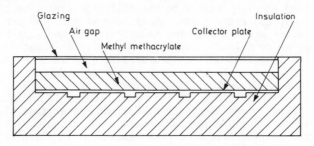

Fig. 3.9. Thermal Trap Collector

Honeycomb systems. The use of honeycomb cellular structures placed
between the transparent cover and the collector plate has been shown to be an
effective method for improving the overall collector performance by
suppressing natural convection losses and greatly reducing the infra-red
reradiation losses. The cellular material should have a low thermal conductivity to reduce conduction heat losses from the collector plate to the
outer cover. Theoretical studies (47) have predicted that a thin-film transparent plastic honeycomb could increase collector efficiency to greater than
60% at a mean collector temperature of 365 K compared with a measured 43%
with a conventional double-glazed collector with a selective surface. It was
considered that this could be achieved without an increase in collector cost
as only one transparent cover is required for a honeycomb system. Tests
carried out on a polyethylene square-cell array (48), which had 25.4 mm
square sides and a depth of 76.2 mm, showed that natural convection losses
had been effectively suppressed with the collector and the honeycomb in an
inclined position. Earlier work had been confined to horizontal testing.

The University of California, Los Angeles, are strong supporters of glass
as the cellular structure material (49,50) as it is inexpensive and readily
available with low thermal conductivity. Its optical properties are
excellent, as it has a very low solar absorptance and the transmitted and
reflected components of direct solar radiation are specular, allowing the
radiation to maintain its direction towards the collector plate. For a
cellular structure consisting of an array of circular tubes the main design
parameters are the internal diameter, which must be less than 150 mm, and the
length, which should be less than four times the diameter. Other cellular
materials under consideration have reflecting surfaces, but if these are
metallised, the coating must be very thin to reduce the conductive heat loss.

Structurally integrated collectors. In any new installation, or in an
existing building where the roof has to be replaced, considerable economic
advantages can be obtained if the solar collector is combined into a
structural unit so that the collector is also the roof of the building. The
design criteria established by the Los Alamos Scientific Laboratory (51)
included good thermal performance, economical in large scale production, the
use of economical and readily available materials, a long service life and
capable of being easily installed and maintained by local builders. The main
features of the collector are shown in Fig. 3.10. The collector surface was
formed by welding thin mild steel plates with seam welds around the periphery
and central spot welds. The plates were then expanded under pressure to form
the flow channels. The lower extended surface of the collector plate has
three bends so that it forms a structural channel. The upper extended surface is positioned at right angles to the collector plate, so that adjacent
units can be easily connected at the upper edge with a U-shaped cap strip.
The two glass covers are set into the trough section and supported at the
edges by a neoprene or silastic support. Each unit is about 610 mm wide and
from 2.4 to 6.1 m long. As well as the structural advantages already
mentioned, the cap strip with its compression seal reduces the time which
would otherwise be spent on sealing manually on the site. The foam insulation also increases the rigidity of the panel. Glass is the cover material
and among the reasons given for the use of glass rather than plastic are the
problems of sealing and expansion compensation which could arise with the
relatively higher coefficients of thermal expansion in plastics materials. A
full series of test results, including the effects of atmospheric corrosion,
internal corrosion and materials stability were presented at Los Angeles in

1975 (52).

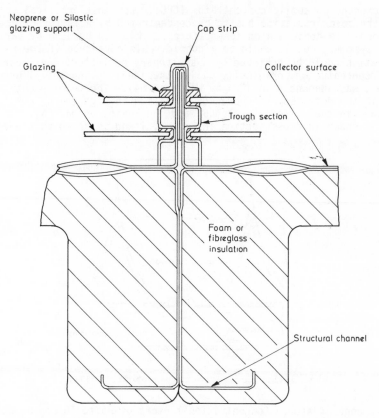

Fig. 3.10.   Structurally Integrated Collector

Negative pressure distributed flow collectors. A collector design which
has overcome the necessity for the water channels to withstand a positive
internal pressure has been developed at the University of Iowa (53).   The
design concept is based on flow between two parallel plates and uses either
corrugations or other forms of surface indentations on one or both sheets, or
a porous spacer, such as a screen wire, between the two sheets. Flow through
the collector is downward and at sub-atmospheric pressure, the whole of the
under surface of the collector plate being exposed to the heated water. A
significant improvement in collector performance was shown when compared with
some commercially available conventional collectors. At an estimated
temperature difference of 52.5°C above the surroundings with an incident
radiation of 750 W/m², the negative pressure distributed flow collector had
an overall efficiency of 44% compared with 38.4% from a conventional
collector. The material for the distributed flow collector plate could
probably be copper sheet of 1.27 mm thickness, which would give adequate
strength to withstand the compressive force caused by the atmospheric
pressure-fluid pressure differential. The use of such thin sheet copper considerably reduces the estimates materials cost of these collectors in
production when compared with conventional collectors.

Some Non-Tracking Reflecting and Concentrating Systems

The compound parabolic concentrator (CPC). For most practical solar power
schemes the solar radiation has to be concentrated by a factor of about 10 or
more in order to achieve high temperatures. This can be done by various
tracking systems, but it would be a considerable advantage if the required
concentration could be achieved by a stationary collector. An important
class of concentrator, originally called the ideal cylindrical light
collector, was announced in 1974 by Winston (54). The development had its
origins in the detection of Cherenkov radiation in high energy physics
experiments in the United States (55) and the Soviet Union (56). The basic
concept is shown in cross-section in Fig. 3.11 and is known as the compound

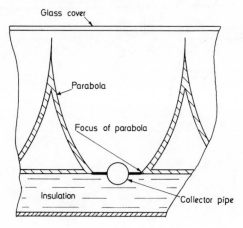

Fig. 3.11. Concentrating Parabolic Collector

parabolic concentrator. Concentration factors of up to 10 can be achieved
without diurnal tracking and if lower factors, in the order of 3, are
acceptable, even seasonal adjustments may not be needed. The efficiency for
accepting diffuse radiation is much larger than for conventional focussing
collectors and is the reciprocal of the concentration factor. As shown in
Fig. 3.11, the focus of the right-hand parabola is at the base of the left-
hand parabola, and vice-versa. The axis of each parabola is inclined to the
vertical optic axis. Heat collection can be achieved by adding a cylindrical
black body collector at the base of the parabolic array or by extending the
parabola to encircle any particular collector shape. A review of some of
these alternatives has been given by Rabl (57).

The spiral or 'sea shell' collector. An extension of the compound parabolic concentrator into a single-sided parabolic section ending with a
circular reflector was also described by Rabl (57). As shown in Fig. 3.12, a
spiral collector contains an inwardly spiralling curved section. When direct
radiation enters the spiral it cannot be reflected outwards and continues to
be reflected deeper into the spiral until the absorbing section is reached,
shown as a circular cross-section in Fig. 3.12. For the solar thermal
generation of electricity Smith (58) suggests a parabolic entry section
followed by a spiral curve and evacuating the space around the collector. A
patent application for a 'mathematically shaped reflector such that all
reflectable radiation, diffuse or specular, entering the device through the

aperture must ultimately strike the collecting member and cannot be reflected outwards' has been filed in Australia (59) while one of the prize-winning entries in the UK Copper Development Association's Solar Heating Competition in 1975 featured a logarithmic or equiangular spiral system (60).

The trapezoidal moderately concentrating collector. Moderate concentration of solar energy can be achieved by redirecting the incident radiation on a given area onto a smaller area. Since focussing is not requires, both

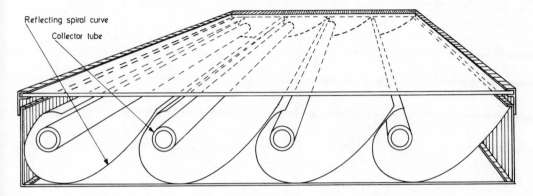

Fig. 3.12.

direct and diffuse radiation can be used. A simple, easily constructed collector of this type consists of a series of long, trapezoidal, parallel stationary grooves, as shown in cross-section in Fig. 3.13. The grooves have

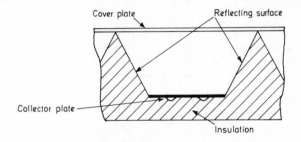

Fig. 3.13.

highly reflective sidewalls and the base is the absorbing surface of the collector. Because the absorbing surface is smaller than the overall collector area, longwave emission losses are reduced. The term 'directional selectivity' can be used in describing this feature and it was demonstrated for a vee-groove configuration by Hollands (61). Results presented in 1975 by Bannerot and Howell (62,63) were used to produce design nomographs for different geometries and indicate that this class of collector could have considerable potential for applications in commercially available absorption cooling, where flat plate collectors have reached the limit of their useful output at about 100 to 150°C.

The Stationary Reflecting/Tracking Absorber (SRTA)

The collector, which is shown in Fig. 3.14, consists of a segment of a
spherical mirror placed in a stationary position facing the sun. It has a
linear absorber which can track the image of the sun by a simple pivoting
motion about the centre of curvature of the reflector (64,65). From

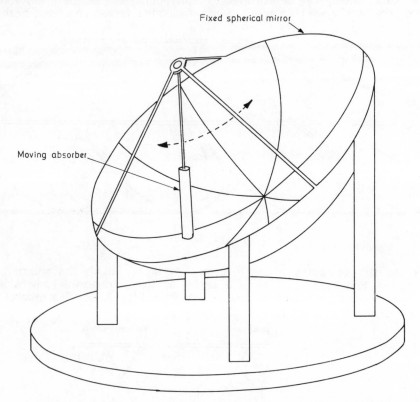

Fixed spherical mirror

Moving absorber

Fig. 3.14.   The Stationary Reflecting/Tracking
             Absorber (SRTA)

experience gained with pre-production prototypes it has been estimated that
for large scale applications in the range from 10 to 100 MW it would be
feasible to produce electric solar power at less than the projected costs for
nuclear power. Among the advantages mentioned for the SRTA in building
applications are that the same system can be used for electric power and hot
water, the working fluid can reach a high temperature, thus reducing the size
of the storage system, and that there is no danger of breakages with large
glazed areas. The main disadvantage in countries with a high proportion of
diffuse radiation is that it is designed to absorb direct radiation and will
not absorb much diffuse radiation. It has very considerable potential in
countries with a high proportion of direct radiation for various small-scale
power applications.

Evacuated Systems

An alternative approach to the problem of reducing heat losses from flat
plate collectors at high temperatures, typically in the range from 80 to
150°C, is to use an evacuated collector. The combined use of a moderate
vacuum, 1 mm Hg (133.3 N/m^2 or 1 torr), and conventional selective absorber
surfaces in Dallas, Texas (66,67) showed that it was possible to operate at
a temperature of 150°C with a daily energy collection efficiency of more than
40%. The spacing between the absorber surface and the glass cover was found
to be very critical in suppressing the natural convection and conduction
losses. There are various practical difficulties with this system, but they
are not considered to be formidable. Early tests experienced trouble with

Fig. 3.15.

seals, but a technique of using high temperature silicone sealants has been developed. Acrylic covers used in the early tests have been replaced by tempered or chemically-strengthened glass covers. This system was developed into a pre-production commercial prototype in 1975 (68).

Several commercial groups have developed evacuated tubular collectors (24, 69, 70) and the basic module of the Owens-Illinois collector, first demonstrated in 1975, is illustrated in Fig. 3.15. Each module contains 24 tubes, nominally 50 mm diameter x 1.12 m long and several major commercial systems had been installed by 1975, including a 46.5 m^2 system for cooling in Los Angeles and a 93 m^2 system in an office building in Detroit. A cross-section through a tube is shown in Fig. 3.16 compared with the Philips type. The Owens-Illinois tube has a vacuum pressure of less than 10^{-4} mm Hg, a cover tube transmittance of 0.92 and a selective coating applied to the outer surface of the absorber tube (τ = 0.86 and e = 0.07). The absorber tube is supported at its free end by a spring clip to allow for differential expansion and is hermetically sealed to the cover tube. The feeder tube provides the reverse flow path for the heat transfer fluid. An inexpensive diffuse reflecting surface behind the tubes has been shown to nearly double the energy intercepted by the tube (69). The Philips collector was a double absorber tube system with the outer cover having an internal transparent selective coating on its upper half and a reflecting mirror surface on its lower half. The transparent selective coating of indium oxide, In_2O_3, has a

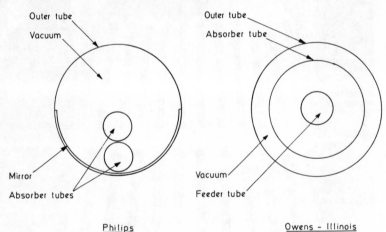

Fig. 3.16. Evacuated Tubular Collectors

transmittance τ of 0.85 and a reflectance ρ of about 0.9 in the range of infra-red radiation from the absorber tubes between 300 and 400 K. The Philips group have emphasized the good performance which can be obtained with this system under the generally diffuse radiation conditions of Northern Europe. Confirmation of the need for a vacuum pressure less than 10^{-4} mm Hg with this type of collector was given by work reported from Australia (71), where moderate vacuums in the order of 0.5 mm Hg produced no improvement in the performance.

An important feature common to all tubular collectors is that the reflection losses with direct radiation will be much lower compared with a flat

glazed surface. This enables more use to be made of early morning and late
afternoon direct radiation.

The Heat Pipe Collector

The basic components of a heat pipe are shown in Fig. 3.17. A small
amount of liquid, which is in equilibrium with its saturated vapour is sealed
within the pipe (although other configurations may be used). When heat is
applied at one end, evaporation takes place and the excess vapour is condensed at the other, unheated, end of the heat pipe. The condensate is
returned to the heated end by means of capillary forces in the wick section.
In some solar heating applications, the return of the condensate can be
simply achieved by gravity flow. As the process of evaporation and condensation at constant pressure is also a constant temperature operation, the heat
pipe is capable of transferring thermal energy internally with very small

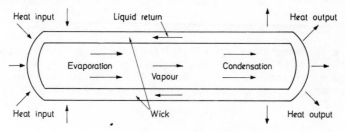

Fig. 3.17. Heat Pipe Collector

temperature differences. There is an inevitable loss of efficiency in transferring the heat from the heat pipe to the secondary circuit. A major
programme of heat pipe collection research has been carried out in the United
States since 1974 (72) and work in Holland during 1975 was reported by
Francken (73), who emphasized the fast thermal response characteristics to
changes in solar radiation. Another advantage is that it can contain fluids
with lower freezing points than water. A heat pipe collector also featured
in the UK Copper Development Association's Competition (60). The preliminary
performance curve shown by a commercial manufacturer in the UK (74) was
rather disappointing, as its overall performance efficiency was slightly
worse than the single glazed non-selective flat plate collector tested by
Heywood (3) in the 1950's.

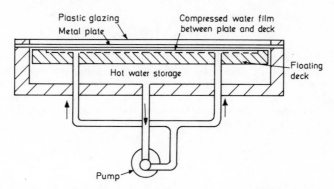

Fig. 3.18. The Floating-Deck Heater

The Floating-Deck Heater

The philosophy behind this development ('75) is that considering the diffuse nature of solar energy, there is as much to be gained by simplicity, low cost and ease of installation as in increased efficiency, although the experimental evidence showed that its efficiency was comparable with other horizontal flat plate collectors. The basic principles are shown in Fig. 3.18. The floating deck is an insulating layer, preferably foam glass, which floats on a hot water storage section. Solar energy is collected by the water flowing in a thin film over the top of the insulation. A compressing layer, which can be transparent or black glass, plastic or metal, rests directly upon the floating water film. After initial tests on a square heater with an area of 0.836 m², a large scale version with an area of 46.5 m² has been successfully developed. It was considered to be suitable for fairly low temperature applications in the lower latitudes, or possibly in combination with a long term storage facility at higher latitudes.

The Cylindrical Heater/Storage System

This is a self-contained cylindrical heat collector and storage vessel developed over several years in New Zealand by Vincze (76,77). The operating principles are illustrated in Fig. 3.19. As the solar radiation reaches the black collector surface, the water in the narrow annular space rises and the

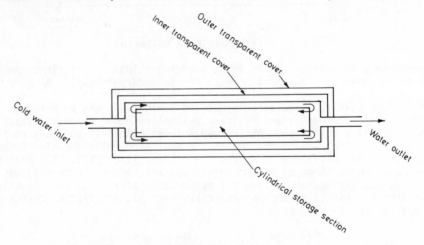

Fig. 3.19.

cooler water inside the vessel descends, establishing a natural thermosyphon. Recent test results (77) indicate that it has a superior performance when compared with a flat plate collector, provided the actual area of the projected cylinder is used as the basis for comparison. If the area needed to space the cylinders apart is taken into account, then the flat plate and the cylindrical have very similar efficiencies.

Air Heaters

Solar air heaters have not attracted nearly as much research and development work as water heating systems (7,78), but there are many applications where air is a more appropriate heat transfer fluid, for example crop drying in the lower latitudes or space heating in the higher latitudes. Air heaters have three particular advantages:-

  (i)   Air cannot freeze;

  (ii) While air can leak, it is not nearly so serious as water leakage;

  (iii) Problems of corrosion in mixed metal systems and storage tanks are less likely.

However, the physical properties of air are less favourable, particularly its relatively very low density and specific heat, and the ducts needed in air systems are very much larger than water pipes.

   Simple air heaters can be made from almost any surface which can be painted black. The three main types of simple heater are shown in Fig. 3.20,

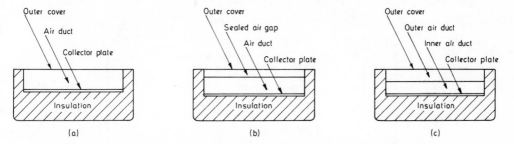

Fig. 3.20. Air Heaters

with single cover plates. In type (a), the duct is the space between the transparent cover and the collector absorber plate. With type (b), there is a sealed air gap between the cover plate and the collector plate to reduce convective heat exchange, and the duct is behind the collector plate. Type (c) uses either a divided air stream or takes the air flow through the outer section as a preheater before passing through the inner air duct. An excellent example of a solar air heater made our of simple materials is an installation with over 500 m^2 of collector in Gujarat (79), where the air is passed through black painted swarf, the metal scrap discarded after metal cutting processes. The double glazed collector has an estimated efficiency of about 45% with a temperature difference of 65ºC above ambient. The early work by Lof in the United States (80) was carried out with a series of over-lapping black glass plates installed under one, two or three cover plates. Lof subsequently installed an overlapped glass plate system in the Colorado Solar House and its performance in the 1959-60 heating season has been reported (81). After 16 years of practically trouble free operation, the system has been investigated again, forming part of the 1976-7 US programme (82). Australia has been another centre for major solar air heater research and the use of selective surfaces in vee-corrugated surfaces was first studied there in the early 1960's (61,83). The air heater shown in Fig. 3.2 (i) is based on this principle.

   As well as selective surfaces, methods of improving collector efficiency include control of the air velocity and the use of a two-pass system (84), in which the air passes between the two glass cover plates of an otherwise conventional double-glazed unit. This particular system showed an effective increase in efficiency of up to 17% compared with operation as a conventional unit. Other systems use various types of finned surface to improve heat transfer (85) as shown in Fig. 3.2 (h), and the use of honeycombs combined with a porous bed has also been tried (86).

The integration of an air collector into a heating and cooling system is shown in Fig. 3.21, which is based on diagrams given by Lof (78). The

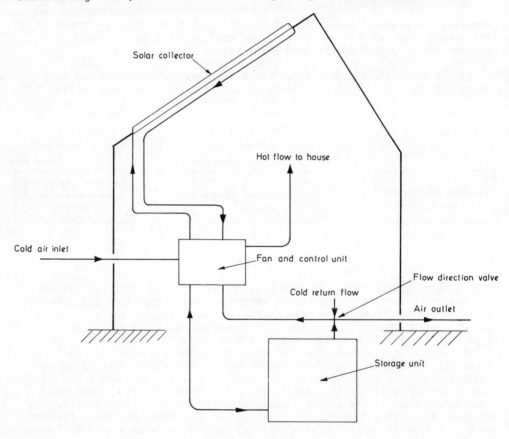

Fig. 3.21.

storage unit is of the 'rock storage' type and in this case consists of ordinary screened gravel. The fan and control unit can direct the air flow in any of the following modes:-

    (i)   Direct heating of the house from the collector;

    (ii) House heating from the storage unit;

    (iii) Storage of heat from the collector;

(iv) Cooling the storage unit from external cold air;

(v) House cooling from the storage unit.

The dual use of the storage unit for both summer cooling and winter heating
is an added attraction. The auxiliary heating system is omitted from the
diagram.

Comparative Performance of Collectors

The overall thermal efficiency of a collector has been derived from the
Hottel-Whillier-Bliss equation in equation 3.2, and this can be used to compare the performance characteristics of different types of collector.
Figure 3.22 shows the efficiency plotted against $(T_m - T_a) \times G_c^{-1}$ for four-
teen different types of collector, based on data available in 1976. It is
probable that this type of graph will become increasingly important as
national standards of performance become established throughout the world.

The characteristic curves of all but the two simplest types, the low
temperature rise (40) and the trickle (46,87), pass through the rectangle
bounded by 50 to 70% efficiency and 0.03 to 0.05 $(T_m - T_a) \times G_c^{-1}$. This
means that for incident radiation intensities greater than 500 W/m^2 all these
heaters will have similar performances for temperature rises between 15 and
30^oC above ambient, the most common range for domestic water heating applications. The trickle type and the low temperature rise type are not suitable
for high temperature applications, as the maximum possible rise appears to be
about 60^oC at zero efficiency. A commercial flat plate collector with a good
performance characteristic, the Honeywell double-glazed collector with an
antireflection (AR) surface on the glazing and a selective surface on the
collector absorber plate (88), has a very similar performance to the con-
ceptually simpler thermal trap system (46), although the latter had still to
be evaluated under field conditions. The compound parabolic concentrator
(CPC) (89) and the heat pipe collector (74) can be expected to have
significantly improved performance characteristics towards the end of the
1970's, as these results are among the first to be reported. Both collectors
are the subject of intense research programmes in the United States. The two
evacuated tubular collectors (24,69) perform very well at high temperature
differentials under good radiation conditions, but also have a good
performance under poor radiation conditions. Heywood's early results for
double and single glazing (3) can be used as a standard for all simple flat
plate collectors, while the PPG collector (90) is representative of a more
advanced commercially available type. Simple air heaters (3, 81) have a
relatively good performance when compared with conventional water heaters.

Although Fig. 3.22 is based on an equation with many simplifying
assumptions, it enables a fair comparison to be made between collectors
tested in different locations under very different radiation conditions. It
does not give an economic assessment or comparison and collectors with very
similar thermal performance characteristics could differ by a factor of at
least two in their cost. The estimated service life is another important
practical feature which cannot be evaluated in this analysis.

Test Procedures

With an increasing number of new solar heaters appearing in the late

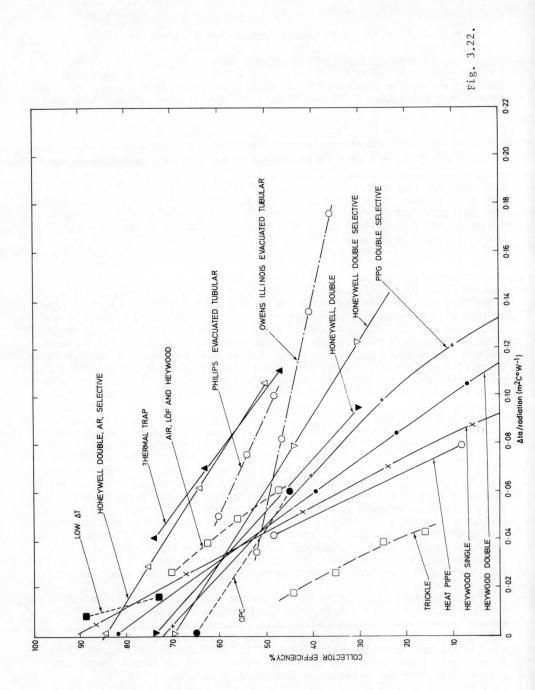

Fig. 3.22.

1970's, it is very important that an internationally agreed standard testing
procedure can be established. The first country to establish a national
standard was Israel in 1966 (91), following earlier work at the National
Physical Laboratory of Israel (92), and a more recent paper from Israel by
Tabor (93) outlines a testing procedure using a governing equation which is
essentially equation 3.2, the modified Hottel-Whillier-Bliss equation.
Tabor's suggested experimental procedure involves the connexion of up to four
similar collectors in series. At any instant of time the solar radiation
intensity and flow rates are identical for each collector and a number of
points on the performance characteristic curve, similar to Fig. 3.22, can be
obtained from one test. The procedure also calls for testing on a clear day
with little wind and estimates for a typical flat plate collector show the
efficiency for values of $(T_m - T_a)$ = 4.5°C to drop about 0.5% for a wind
change from zero to 4.47 m/s, 6.5% at $(T_m - T_a)$ = 26.7°C and 19.5% at
$(T_m - T_a)$ = 48.9°C. A non-linear theoretical model to allow for the effect
of wind velocity has been suggested in Australia (94).

A draft standard for both solar collectors (95) and thermal storage
devices (96) has been published in the United States. This also uses the
concept of equation 3.2 and defines in considerable detail the methods of
measuring the various parameters, such as temperature, pressure, flow rate
and radiation intensity. At least four test points at recommended values of
10, 30, 50 and 70°C for $(T_m - T_a)$ are suggested to establish the characteristic curve of any collector. The use of a solar simulator, or artificial
sun, enables a collector to be tested under standard conditions of ambient
temperature, wind and radiation intensity and performance results obtained
with a simulator in the United States (97,98) have been found to be in good
agreement with outdoor performance results. A solar simulator facility forms
part of the research programme at the University of Cardiff in the UK. In
countries with variable radiation conditions, solar simulators could play an
important part in testing procedures.

Thermal Energy Storage

Thermal energy storage is essential for both domestic water and space-
heating applications and for the high temperature storage systems needed for
thermal power applications. There are other applications, such as horticultural or in the process industries, where a storage facility is also
required. For certain applications, particularly the need for space-cooling
during summer months, it is an advantage if the storage unit can also store
at low temperatures. The choice of storage material depends on the
particular application and for many domestic applications water and/or rock
storage systems have been developed. A combined solar air heater/rock
storage structure is illustrated in Fig. 3.23. First described in 1974 (99),
it is a fully portable A-framed insulated unit containing washed river gravel
with the air heater on the south facing sloping wall assisted by a hinged
reflecting surface, which can be used to cover the collector at night to
prevent heat losses. Studies on the use of gravel bed storage have been
carried out for several years in Australia (100,101,102) and the advantages
of replacing gravel with a strongly adsorbing material such as silica gel or
activated alumina are discussed by Close and Pryor (103).

Water and rocks are typical examples of materials which store energy as
specific heat (sensible heat), but their use is limited by their comparatively low specific heats. The heat of fusion (latent heat) which is involved

Fig. 3.23.

TABLE 3.3

Thermal storage of 1 GJ

	Rock	Water	Heat of fusion materials
Specific heat, kJ/kg °C	0.837	4.187	2.09
Heat of fusion, kJ/kg	-	-	232.6
Density, kg/m³	2242	1000	1602
Storage of 1 GJ, weight, kg	59737	11941	3644
Relative weight	16.4	3.27	1
Volume, m³	26.6	11.941	2.274
Relative volume	11.69	5.25	1

when a substance changes state from a solid to a liquid provides an attractive method of storing a given amount of heat within a much smaller volume. This is illustrated in Table 3.3, which shows the weight and volume of rock-like solids, water and heat of fusion materials for storing 1 GJ (about 278 kWh) with a 20°C temperature limit. The table is based on data originally given by Telkes (104), who discussed the properties of a wide

Fig. 3.24.

range of salt hydrates which could be used for heat storage. The least
expensive and most readily available is sodium sulphate decahydrate
($Na_2SO_4.10H_2O$) or Glauber's Salt, which has to be mixed with 3 to 4% Borax
as a nucleating agent if complete crystallization is to be obtained. These
processes occur at about $30^{o}C$. For high temperature storage, in the order of
200 to $300^{o}C$, other salts have been considered (105,106) and the heat of
hydration of inorganic oxides, principally MgO and CaO (107). The thermal

Fig. 3.25.

interaction between any underground storage system and the surrounding ground has also been studied analytically (108).

The use of the space-heated house as the energy store is well-known, but an original idea in conserving heat inside the house is the Zomework's bead wall, developed by Steven Baer (109). This is illustrated in Fig. 3.24 (emptying) and Fig. 3.25 (filling). Styrofoam beads are automatically blown between two panes of glass to prevent heat losses at night in the winter months or they can be used to stop unwanted heat gains in the summer. The system has an advantage compared with folding doors or shutters which require free space next to the window.

References

(1) Brooks, F.A., Solar energy and its uses for heating water in California, Bull. Calif. Agric. Exp. Sta., No. 602, 1936.

(2) Hottel, H.C. and Woertz, B.B., The performance of flat plate solar heat collectors, Trans. ASME, 64, 91-104, 1942.

(3) Heywood, H., Solar energy for water and space heating, J. Inst. Fuel 27, 334-347, July 1954.

(4) Morse, R.N., Solar water heaters, Proc. World Symposium on Applied Solar Energy, Stanford Research Inst., University of Arizona, Phoenix, Arizona, 191-202, 1956.

(5) Chinnery, D.N.W., Solar water heating in South Africa, National Building Research Institute, Bulletin 44, CSIR Research Report 248, Pretoria, South Africa, 1967.

(6) Tabor, H., Solar energy collector design, Bull. Res. Coun., 5C, No. 1, Israel, 1955.

(7) Yellott, J.I., Solar energy utilization for heating and cooling, originally published in ASHRAE Journal, December 1973, now in Chapter 59, ASHRAE Guide and Data Book series, 1974 edition.

(8) Scott, J.E., User's experience with solar water heater collectors in Florida, Proc. Workshop on Solar Collectors for Heating and Cooling of Buildings, 21-23 November 1974, NSF-RANN-75-019, May 1975.

(9) CSIRO Solar Water Heaters, Division of Mechanical Engineering Circular No. 2, Melbourne, 1964.

(10) McVeigh, J.C., Some experiments with a flat plate solar water heater, UK Section, ISES, Conf. on Low Temperature Thermal Collection of Solar Energy, April 1974.

(11) McVeigh, J.C., Low-cost solar water heater, Proc. Conf. on Appropriate Technology, University of Newcastle-upon-Tyne, 1976.

(12) Whillier, A., Solar energy collection and its utilization for house
      heating, ScD. Thesis, MIT, 1953.

(13) Whillier, A., Design factors influencing collector performance, Low
      Temperature Engineering Applications of Solar Energy, ASHRAE, New
      York, 1967.

(14) Hottel, H.C. and Whillier, A., Evaluation of flat plate collector
      performance, Trans. Conf. on the Use of Solar Energy, 2 (1), 74,
      University of Arizona Press, 1958.

(15) Bliss, R.W., The derivation of several 'plate efficiency factors', useful
      in the design of flat plate solar heat collectors, Solar Energy 3,
      (4), 55 (1959).

(16) Duffie, J.A. and Beckman, W.A., Solar Energy Thermal Process, John
      Wiley & Sons, New York, 1974.

(17) Smith, C.T. and Weiss, T.A., Design applications os the Hottel-
      Whillier-Bliss equation, ISES Congress, Los Angeles, Extended
      Abstracts, Paper 34/6, July 1975.

(18) Mitalas, G.P. and Stephenson, D.G., Absorption and Transmission of
      Thermal Radiation by Single and Double Glazed Windows, Research
      paper 173, Division of Building Research, National Research Council
      of Canada, Ottawa, 1962.

(19) Zarem, A.M. and Erway, D.D., Introduction to the Utilization of Solar
      Energy, McGraw Hill, New York, 1963.

(20) Scoville, A.E., An alternate cover material for solar collectors, ISES
      Congress, Los Angeles, Extended Abstracts, Paper 30/11, July 1975.

(21) Charters, W.W.S. and Macdonald, R.W.G., Heat transfer effects in solar
      air heaters, Paper E 37, Conf. The Sun in the Service of Mankind,
      UNESCO, Paris, 1973.

(22) Lorsch, H.G., Performance of flat plate collectors, Proc. Solar Heating
      and Cooling for Buildings Workshop, 21-23 March 1973, NSF-RANN-73-
      004, July 1973.

(23) Tani, T., Sawata, S., Tanaka, T. and Horigome, T., Characteristics of
      selective thin barriers and selective surfaces, ISES Congress, Los
      Angeles, Extended Abstracts, Paper 30/4, July 1975.

(24) Bruno, R., Herman, W., Horster, K., Kersten, R. and Mahdjuri, R., High
      efficiency solar collectors, Paper 34/10, Ibid.

(25) McDonald G.E., Spectral reflectance properties of plated zinc for use
      as a solar selective coating, Paper 30/2, Ibid.

(26) Solar Energy : a UK Assessment, UK Section, ISES, London, May 1976.

(27) Selective Black Coatings, Proc. UN Conf. on New Sources of Energy, 4,
      618, 1964.

(28) Keller, A., Selective surfaces of copper foils, Paper E 43, Conf. The
      Sun in the Service of Mankind, UNESCO, Paris, 1973.

(29) Keller, A., Selective surfaces of aluminium foils, Paper 7/16, Inter-
      national Solar Energy Society Conference, Melbourne, 1970.

(30) Close, D.J., Flat plate solar absorbers. The production and testing of
      a selective surface for copper absorber plates. Report ED 7, CSIRO,
      Melbourne, June 1962.

(31) McDonald, G.E., Refinement in Black Chrome for use as a solar selective
      coating, NASA TM X-3136, 1975.

(32) Pettit, R.B. and Sowell, R.R., Solar absorptance and emittance
      properties of several solar coatings, ISES Congress, Los Angeles,
      Extended Abstracts, Paper 30/1, July 1975.

(33) Harris, L., The optical properties of metal blacks and carbon blacks,
      The Eppley Foundation for Research Monograph Series No. 1, 1967.

(34) McKenzie, D.R., Harding, G.L. and Window, B., Metal blacks as selective
      surfaces, ISES Congress, Los Angeles, Extended Abstracts, Paper 30/5,
      July 1975.

(35) Sabine, T.M., Gammon, R.B. and Riddiford, C.L., "Solarox" as a select-
      ive absorber, Paper 30/6, Ibid.

(36) Stamford, M.S., Compatibility of solar systems, A question of solar
      heating, Copper Development Association, Potters Bar, Hertfordshire,
      1976.

(37) Private communication, Department of Mechanical Engineering and Energy
      Studies, University of Cardiff, 1976.

(38) Duffie, J.A. et al, Report of working group on materials and components
      for flat plate collectors, Proc. Workshop on Solar Collectors for
      Heating and Cooling of Buildings, 21-23 November, 1974, NSF-RANN-75-
      019, May 1975.

(39) Popplewell, J.M., Corrosion considerations in the use of aluminium and
      copper solar energy collectors, ISES Congress, Los Angeles, Extended
      Abstracts, Paper 30/12, July 1975.

(40) McVeigh, J.C., Some experiments in heating swimming pools by solar
      energy, JIHVE 39, 53-55, June 1971.

(41) Czarnecki, J.T., Method of heating swimming pools by solar energy,
      Solar Energy 7 (1), 3-7 (1963).

(42) Root, D.E., Practical aspects of swimming pool heating, Solar Energy
      4 (1), 23-24 (1960).

(43) Farber, E.A. and Triandafyllis, J., Solar swimming pool heating, Conf.
      The Sun in the Service of Mankind, UNESCO, Paris, 1973.

(44) Farber, E.A., Solar energy research and development at the University
        of Florida, Building Systems Design, February/March, 1974.

(45) Cobble, M.H., Irradiation into transparent solids and the thermal trap
        effect, J. Frank. Inst. 278 (6), 383-393 (1964).

(46) San Martin, R.L. and Fjeld, G.J., Experimental performance of three
        solar collectors, Solar Energy 17 (6), 345-349 (1975).

(47) Marshall, K.N., Bell, G.A., Wedel, R.K. and Haslim, L.A., Thermal
        radiation characteristics of transparent plastic honeycombs for
        solar collector applications, ISES Congress, Los Angeles, Extended
        Abstracts, Paper 32/1, July 1975.

(48) Cane, R.L.D., Hollands, K.G.T., Raithby, G.D. and Unny, T.E.,
        Convection suppression in inclined honeycombs, Paper 32/5, Ibid.

(49) Buchberg, H., Edwards, D.K. and Mackenzie, J.D., Design considerations
        for solar collectors with glass cylindrical cellular covers, Paper
        32/12, Ibid.

(50) Baldwin, C.M., Dunn, B.S., Hilliard, W.G. and Mackenzie, J.D.,
        Performance of transparent glass honeycombs in flat plate collectors,
        Paper 32/2, Ibid.

(51) Moore, S.W., Balcomb, J.D. and Hedstrom, J.C., Design and testing of a
        structurally integrated steel solar collector unit based on
        expanded flat metal plates, Presented at US Section ISES Meeting,
        Ft. Collins, Colorado, 19-23 August, 1974.

(52) Balcomb, J.D., Hedstrom, J.C. and Moore, S.W., The LASL structurally
        integrated solar collector unit - final results, ISES Congress,
        Los Angeles, Extended Abstracts, Paper 34/1, July 1975.

(53) Spencer, D.L., Smith, T.F. and Flindt, H.R., The design and performance
        of a distributed flow water-cooled solar collector, College of
        Engineering, University of Iowa, Iowa 52242, 1975.

(54) Winston, R., Principles of solar concentrators of a novel design,
        Solar Energy 16 (2), 89-95 (1974).

(55) Hinterberger, H. and Winston, R., Rev. Sci. Instr. 37, 1094 (1966).

(56) Varanov, V.K. and Melnikov, G.K., Soviet Journal of Optical Technology
        33, 408 (1966).

(57) Rabl, A., Comparison of solar concentrators, Solar Energy 18 (2),
        93-111 (1976).

(58) Smith, R.H., A method of solar thermal generation of electricity,
        ISES Congress, Los Angeles, Extended Abstracts, Paper 53/8, July
        1975.

(59) Ling, M., Solarhot Water Systems, 34 Flinders Road, Earlwood, Sydney,
        N.S.W. 2206.

(60) Solar Heating Competition, Copper Development Association, Orchard
      House, Mutton Lane, Potters Bar, Hertfordshire EN6 3AP, UK (1975).

(61) Hollands, K.G.T., Directional selectivity, emittance and absorptance
      properties of vee corrugated specular surfaces, Solar Energy 7 (3),
      108-116 (1963).

(62) Bannerot, R.B. and Howell, J.R., Moderately concentrating flat plate
      solar energy collectors, ASME paper 75-HT-54, presented at AIChE-
      ASME Heat Transfer Conference, San Francisco, California,
      11-13 August 1975.

(63) Bannerot, R.B. and Howell, J.R., The effect of non-direct insolation on
      the radiative performance of trapezoidal grooves used as solar
      energy collectors, ISES Congress, Los Angeles, Extended Abstracts,
      Paper 52/5, July 1975.

(64) Steward, W.G., A concentrating solar energy system employing a
      stationary spherical mirror and movable collector, Proc. Solar
      Heating and Cooling for Buildings Workshop, 21-23 March 1973, NSF-
      RANN-73-004, July 1973.

(65) Kreider, J.F., The Stationary Reflector/Tracking Absorber Solar
      Collector, Presented at US Section ISES Meeting, Ft. Collins,
      Colorado, 19-23 August 1974.

(66) Blum, H.A. and Estes, J.M., Design and feasibility of flat plate solar
      collectors to operate at 100 - 150°C, Paper E 18, Conf. The sun in
      the Service of Mankind, UNESCO, Paris, 1973.

(67) Eaton, C.B. and Blum, H.A., The use of moderate vacuum environments as
      a means of increasing the collection efficiencies and operating
      temperatures of flat plate solar collectors, Solar Energy 17 (3),
      151-158 (1975).

(68) Kittle, P.A. and Cope, S.L., Outside performance of moderate vacuum
      solar collectors, ISES Congress, Los Angeles, Extended Abstracts,
      Paper 32/8, July 1975.

(69) Beekley, D.C. and Mather, J.R., Analysis and experimental tests of high
      performance tubular solar collectors, Paper 32/10, Ibid.

(70) Ortabasi, U. and Buehl, W.M., Analysis and performance of an evacuated
      tubular collector, Paper 32/11, Ibid.

(71) Read, W.R. and Christie, E.A., Thermal characteristics of evacuated
      tubular solar collectors, Paper 32/9, Ibid.

(72) Bienert, W.B., Heat pipes applied to flat plate solar collectors, Proc.
      Workshop on Solar Collectors for Heating and Cooling of Buildings,
      21-23 November 1974, NSF-RANN-75-019, May 1975.

(73) Francken, J.C., The heat pipe fin, a novel design of a planar collector,
      ISES Congress, Los Angeles, Extended Abstracts, Paper 34/8, July
      1975.

(74) Redpoint Associates Ltd., Cheney Manor, Swindon SN2 2QN, Wiltshire, UK. (1976).

(75) Davison, R.R., Harris, W.B. and Chan Ho Kai, Design and performance of the compressed-film floating deck solar water heater, ISES Congress, Los Angeles, Extended Abstracts, Paper 34/9, July 1975.

(76) Vincze, S.A., A high-speed cylindrical solar water heater, Solar Energy 13, 339-344 (1971).

(77) Vincze, S.A., Comparative winter tests, cylindrical versus flat plate solar heat collectors, ISES Congress, Los Angeles, Extended Abstracts, Paper 14/8, July 1975.

(78) Lof, G.O.G., Space heating with solar air collectors, Proc. Workshop on Solar Collectors for Heating and Cooling of Buildings, 21-23 November 1974, NSF-RANN-75-019, May 1975.

(79) Chandran, T.C., Private Communication, Kaira District Co-operative Milk Producers' Union Ltd., Anand. 388001, Gujarat, India, 1976.

(80) Lof, G.O.G. and Nevens, T.D., Heating of air by solar energy, Ohio Journal of Science 53 (5), 272-280 (1953).

(81) Lof, G.O.G., El Wakil, M.M. and Chion, J.P., Design and performance of domestic heating system employing solar-heated air - the Colorado Solar House, Proc. UN Conf. New Sources of Energy, 185-197, 1964.

(82) Ward, J.C., Long term (16 years) performance of an overlapped-glass plate solar-air heater, Proc. Workshop on Solar Collectors for Heating and Cooling of Buildings, 21-23 November 1974, NSF-RANN-75-019, May 1975.

(83) Close, D.J., Solar air heaters for low and moderate temperature application, Solar Energy 7 (3), 117-124 (1963).

(84) Satcunanathan, S. and Deonarine, S., A two-pass solar air heater, Solar Energy 15, 41-49 (1973).

(85) Bevill, V.D. and Brandt, H., A solar energy collector for heating air, Solar Energy 12, 19-29 (1968).

(86) Lalude, O.A. and Buchberg, H., Design and application of honeycomb porous-bed solar air heaters, Solar Energy 13, 223-242 (1971).

(87) Brachi, P., Sun on the roof, New Scientist, 19 September 1974.

(88) Ramsey, J.W. and Borzoni, J.T., Effects of selective coatings on flat plate solar collector performance, ISES Congress, Los Angeles, Extended Abstracts, Paper 34/5, July 1975.

(89) Boyd, D.A., Gajewski, R., Granetz, R.S. and Winston, R., A concentrating flat plate collector, American Science and Engineering Inc., Cambridge, Mass. and Enrico Fermi Institute, University of Chicago, Paper ASE-3739-B, October 1975.

(90) PPG Industries, Baseline Solar Collector, One Gateway Center, Pittsburgh, Pa. 15222, USA. (1974).

(91) Israeli Standard No. 609, Solar Water Heaters : Test Methods, Israeli Standards Institute, Tel Aviv, May 1966.

(92) Doron, B., Testing of solar collectors, Solar Energy 9 (2), 103-4 (1965).

(93) Tabor, H., The testing of solar collectors, The Scientific Research Foundation, Jerusalem, March 1975 and ISES Congress, Los Angeles, Extended Abstracts, Paper 33/8, July 1975.

(94) Dunkle, R.V. and Cooper, P.I., A proposed method for the evaluation of performance parameters of flat plate solar collectors, Paper 33/2, Ibid.

(95) Hill, J.E. and Kusada, T., Method of testing for rating solar collectors based on thermal performance, NBSIR 74-635, National Bureau of Standards, Washington DC, 20234, December 1974.

(96) Kelly, G.E. and Hill, J.E., Method of testing for rating thermal storage devices based on thermal performance, NBSIR 74-634, National Bureau of Standards, Washington DC, 20234, May 1975.

(97) Simon, F.F. and Harlament, P., Flat plate collector performance evaluation : the case for a solar simulator approach, NASA TM X-71427, October 1973.

(98) Simon, F.F., Flat plate solar collector performance evaluation with a solar simulator as a basis for collector selection and performance prediction, ISES Congress, Los Angeles, Extended Abstracts, Paper 33/4, July 1975.

(99) Keyes, J.H., Project Sungazer : A vertical-vaned flat plate collector with forced-air heat transfer, Proc. Workshop on Solar Collectors for Heating and Cooling of Buildings, 21-23 November 1974, NSF-RANN-75-019, May 1975.

(100) Close, D.J. et al. Design and performance of a thermal storage air conditioning system, Mechanical and Chemical Transactions of the Institute of Engineers, Australia, 4 (1), 1968.

(101) Read, W.R., Choda, A. and Copper, P.I., A solar timber kiln, Solar Energy 15 (4), 309-316 (1974).

(102) Dunkle, R.V., Design considerations and performance predictions for an integrated solar air heater and gravel bed thermal store in a dwelling, Australian and New Zealand section ISES, Melbourne, July 1975.

(103) Close, D.J. and Pryor, T.L., The behaviour of adsorbent energy storage beds, ISES Congress, Los Angeles, Extended Abstracts, Paper 31/2, July 1975.

(104) Telkes, M., Solar energy storage, <u>ASHRAE Journal</u>, 38-44, September 1974.

(105) Severson, A.M. and Smith, G.A., Salt thermal energy storage for solar systems, ISES Congress, Extended Abstracts, Paper 31/5, July 1975.

(106) Kosaka, M. and Asakina, M., Discussions on heat storage material at low temperature level, Paper 31/3, Ibid.

(107) Ervin, G., Solar heat storage based on inorganic chemical reactions, Paper 31/6, Ibid.

(108) Shelton, J., Underground storage of heat in solar heating systems, <u>Solar Energy</u> 17 (2), 137-143 (1975).

(109) Harrison, D., Beadwalls, <u>Solar Energy</u> 17 (5), 317-319 (1975).

CHAPTER 4

SPACE HEATING APPLICATIONS

The possibility of at least partially heating a building by solar energy has been demonstrated on many occasions over the past forty years. The criteria originally suggested by Telkes (1) in 1949 for solving the problems of collection, storage and distribution of solar energy have been somewhat modified since then as experience has been gained with an increasing number of installations. In the early work, emphasis was placed on the collection of winter sunshine. This has been broadened to include the very substantial contribution that can be obtained from diffuse radiation. The terms 'efficient', 'economic' and 'simple' used by Telkes in relation to the collector have always been the goal of sound engineering practice and an analysis of many solar installations reveals that very few can satisfy all three.

The problem of storing the solar energy collected in the summer for use the following winter has attracted a tremendous research effort. Although the principle of using a very large, heavily insulated water storage tank buried underneath the building was described by Hottel and Woertz in 1942 (2), their comment that the unit was known to be highly uneconomical had a considerable influence on the direction of storage system research over the next two decades. The effect of latitude and local radiation characteristics are now more widely appreciated. It was originally thought that only a few days storage could be economically viable so that the solar energy received during clear winter days would be made available for successive overcast periods and that this could only occur to any extent in parts of the world where there are appreciable amounts of winter sunshine. However, longer storage periods of up to several months have been achieved in several solar houses and considerable reductions in the total volume of the store are possible by the use of chemical storage methods pioneered by Telkes. She also drew attention to the need for a thermostatically controlled distribution system that is simple to operate with minimum attention and disturbance to the occupants and the need to avoid overheating, especially in the rapidly changing weather of spring and autumn, ensuring that the solar heating system is definitely not heating the house in the summer, although for interseasonal storage the heavily insulated store should reach as high a temperature as possible if water or rock storage systems are used.

The term 'solar house' first became familiar in the United States during the 1930's, when architects began to use large, south-facing windows to let the lower slanting rays of the winter sun penetrate into the back of the room (3). It was noticed, however, that while fuel was saved during the day it was not possible to store the solar energy and that at night and during cloudy days the heat loss was so great that there was a relatively small saving on fuel during the entire heating season. Two identical test houses were built at Purdue University, Lafayette under the direction of Professor F.W. Hutchinson to obtain quantitative evidence of the fuel savings from solar gain (4,5,6). Both houses were fitted with sealed double glazing, the orthodox house had a conventional window area while the solar house had a greater area of glass on the south side: Both houses were heated by

electricity and thermostatically maintained at identical temperatures. The
most surprising result to emerge from those tests was that the solar house
required about 16% more heat than the orthodox house during the December-
January test period. It was obvious that the larger solar windows dissipated
a great deal of heat at night and during overcast days. If the houses had
been lived in it was probable that the effect of drawing heavy curtains at
night could have made an appreciable difference to the result.

The investigation of solar space heating applications developed steadily
from this point. The work at the Massachusetts Institute of Technology,
which was started by the Godfrey L. Cabot bequest and which led to the
building of Solar House No. I in 1940, continued with a series of different
solar houses. Dr. G.O.G. Lof, of the University of Colerado, was the
earliest experimenter with solar air heaters, using a total collector area of
approximately one third of the roof and passing the heated air either directly into the rooms or into a heat storage bin filled with crushed rock. The
capacity of this store was about one full day's heating requirement and
approximately 20% fuel savings resulted during the first season in 1946 (7,8).
Thirty years later by the beginning of 1976 the number of solar heated
buildings which had been built since 1940 or which were under construction
exceeded 200. Shurcliff's survey in March 1975 (9) included details of over
a hundred United States buildings but less than twenty for the rest of the
world. By 1976 the UK total alone was approaching twenty, with increasing
encouragement from official Government sources.

The various houses and buildings described in the following sections have
been chosen to illustrate the historical development of solar heating leading
to many current projects throughout the world. Starting with the United
States, where much of the early work was carried out, applications are given
from several other countries with a major section devoted to the United
Kingdom.

Some Applications in the United States

M.I.T. Solar House No. I (2,9,10,11)
Constructed in 1939 as two rooms, an office and a laboratory, it had an
approximate floor area of 46.5 m^2. The greater part of the roof sloped about
30^0 southward and contained 33.45 m^2 of exposed absorber surface placed in
37.9 m^2 of triple glazed collector. The absorber surface consisted of
blackened copper sheet to which parallel copper tubes were soldered. The
basement was filled with a large hot water storage tank of 65.86 m^3 capacity
with an average insulation thickness of 665 mm. In heat requirements, the
building was designed to simulate a moderately insulated six-roomed house.
It was the first 100% solar heated building, as it could store the summer's
heat for winter use, but was regarded as very uneconomic and was demolished
in 1941.

M.I.T. Solar House No. II (9,10,12)
A single storey laboratory building constructed in 1947, its dimensions were
approximately 4.26 x 13.4 x 2.44 m high with a bank of vertical solar
collectors on the south wall consisting of seven different panels, each just
under 10 m^2 in area. Various types of storage system were tested and in
1947-9 it was converted into No. III.

M.I.T. Solar House No. III (9,10,11)

With same floor dimensions as No. II, it had a roof-mounted, double glazed collector with a similar absorber system as No. I, having an area of 37.2 m^2 at a tilt of 57° to the horizontal. Storage was in a 4.5 m^3 capacity cylindrical tank in the roof space. During the mid-winter four months up to 85% of the space heating was provided by the system and a mean annual value of about 90% was subsequently obtained. It was destroyed by fire in 1955.

M.I.T. Solar House No. IV (9,10,13,14)

Built in 1959, and illustrated in Fig. 4.1, it was considered unique (14) in that it was designed as a solar house to make the fullest use of collected energy, waste as little energy as possible and at the same time meet the comfort and space requirements of modern living. It was a two-storey design containing 134.7 m^2 of usable living area. The south elevation of the house above ground consisted entirely of 59.5 m^2 of solar collector sloping at an angle of 60° to the horizontal. The double glazed collector was redesigned and had the copper tube mechanically clipped onto blackened sheet aluminium, the overall measured absorptivity being 0.97. The 5.7 m^3 water storage tank was heavily insulated. In operation the occupants were careful not to alter their individual ways of living to favour solar heating, so that dish and clothes-washing operations were always carried out at the convenience of the housewife and not only when the sun was shining. During the winter season from September 30th 1959 to March 30th 1960, 44% of the space heating load and 57% of the domestic hot water load was borne by the solar energy system. This was rather less than the predicted performance, but was considered to be due to the severe winter weather conditions experienced that year. Maintenance problems caused the system to be abandoned after two years, the overall performance for the two winters being 48% of the total load supplied by the solar energy system.

The Dover House (3,9,10)

Claimed to be the first house heated entirely by solar energy, the solar system was designed by Dr. Maria Telkes, at that time a research associate at M.I.T. A Boston architect, Eleanor Raymond, designed the house which is illustrated in Fig. 4.2. It was build as a private project on the estate of Amelia Peabody in Dover, Massachusetts and was first occupied on Christmas eve in 1949. The double glazed 66.89 m^2 air heating collector occupied the entire vertical south face of the two storey building at second floor level. Each collector panel consisted of two 3.28 x 1.22 m figured glass panes separated by a 19 mm air gap. The absorber surface was made from standard galvanized steel sheets painted with an ordinary matt black paint. Behind each sheet was a 76 mm air space through which the air could circulate on its way to three heat storage collector bins. These bins contained cans of Glauber's salt - sodium sulphate decahydrate ($Na_2SO_4 10H_2O$) and occupied a total volume of about 13.3 m^3. Dr. Telkes felt that the difficulty of storing heat for long periods using water or rocks lay in finding a large enough space and suggested the use of the heat of fusion, or heat of melting of chemical compounds, Glauber's salt, with a melting point of about 32°C, could store about six or seven times more heat than water on an equal volume basis. Only the ground floor area of 135.3 m^2 was heated and the heat was transferred from storage into rooms by means of small fans which responded to individual thermostats. The original storage capacity was estimated to be about 12 full days mid-winter heating load. During the first year of operation the solar system provided the entire heating load, but the performance deteriorated after this, however, due to stratification and irreversibility

Fig. 4.1. MIT Solar House IV

Fig. 4.2.   The Dover House

in the fusion and freezing of the salt, and auxiliary heat was later required.
The solar heating system was removed after four years when the house was
enlarged, but several very important design features had by then been
established:-

(i)   The effectiveness of the solar air collector, with its simple design
      and dual use as a heat collector and as a wall. This dual-use of the
      collector as a wall or part of the roof was to feature in the great
      majority of subsequent solar house designs.

(ii) The advantage of having controlled temperature zones in different
      parts of the house. This was overlooked in many later designs, but is
      once again being recognised as very important for saving energy.

(iii) The large heat storage capacity in a small volume which could be pro-
      vided by heat-of-fusion salts. The problem of stratification with
      repeated cycling was still to be overcome and has proved to be one of
      the most difficult problems in solar space heating applications.

The U.S. Forest Service Bungalow (Bliss House) (15)

An existing single-storey bungalow at Amado, Arizona, with a floor area of
62.43 m^2 was modified in 1954-5 to include a solar air heating system and
massive rock storage. The collector consisted of four layers of black cotton
cloth, with a 12.5 mm gap between each layer, under single glazing. It had
an area of 29.26 m^2 and was erected close to the bungalow at an angle of 53^0
from the horizontal. The storage system was 65 tons, approximately 36.8 m^3,
of 100 mm diameter rock, also located near the bungalow in an insulated
underground chamber. In operation, the air was circulated from the collector
into storage by means of a fan whenever the radiation conditions were
appropriate. A second fan supplied air to the house on demand, either
directly from the collector or from the storage system. The system provided
all the heat necessary for winter space heating in the bungalow and was
claimed to be the first 100% solar heated home in the U.S.A. For summer
cooling, night air was drawn through a black cloth which covered a separate
horizontal bed. This cooled the air further - about 1^0C - and it was then
passed to the storage system. During the day this cooled air could be
circulated to the bungalow. The system was demolished after just over one
year's successful operation.

    An interesting design feature was that ten average days' storage capacity
was provided. This was more than adequate as one day's winter sun provided
over two days' heat demand. In more northerly latitudes much greater storage
capacity is necessary to compensate for the considerably lower winter radiation levels. At that time the economics were adverse, as the capital costs
were over five times greater than a conventional heating system and it was not
possible to repay the capital costs and interest charges with the amount of
fuel saved - the ratio of capital cost to fuel saved being 50:1. It was also
visually unattractive.

The Bridgers and Paxon Albuquerque Office Building (16,17)

The world's first solar heated office building was built in Albuquerque, New
Mexico, and was first occupied in August 1956. The useful floor area was
approximately 400 m^2 and the building had south facing flat plate collectors
inclined at 60^0 to the horizontal as illustrated in Fig. 4.3. The net
collector area was about 70 m^2 with single glazing and there was a 22.7 m^3

Fig. 4.3. The Bridgers and Paxon Albuquerque Building

underground insulated storage tank. All items of equipment in the building
were standard except for the collector plates, which consisted of aluminium
sheets, 0.476 mm thick, painted with a non-selective black paint and with
38 mm outside diameter copper pipes placed 150 mm apart, soldered to form a
continuous bond on the back side, and containing the heated water. Inside
the building, heat was supplied by passing the warm water from the storage
tank at about 40°C through pipes placed in floor and ceiling panels. A heat
pump was also used when storage water temperatures were not at a high enough
level to satisfy building heat requirements.

The advantages of using a heat pump with a solar collector were clearly
stated in the first report on the performance of the building (16). The heat
pump can be used for cooling in the summer season and this dual role is
attractive. In cold and cloudy weather the storage and collecting temperatures may be allowed to fall very low with the resultant increase in
collector efficiencies and storage capacity. Use of the heat pump also
allows a smaller collector and storage tank to be installed. In the first
season of operation direct solar heating supplied 62.7% of the total heating
requirements while the remaining 37.3% was supplied with the heat pump
operating. It must be emphasized that the major source of heat was from the
solar collectors, even when the heat pump was operated. The amount of energy
required to run the heat pump supplied only 8.2% of the total heating
requirement. It is interesting to note the comment that with the cost of
fuels at that time (1956-7) for heating, the savings in fuel costs did not
justify the necessary first cost expense for solar heating systems for most
localities in the United States. But there would be some areas where the
high fuel costs would make solar heating systems economically attractive.
The system operated in its original form for about six years with only
occasional trouble, for example when draining was incomplete and some frost
damage occurred. There was also deterioration in some rubber pipe hose
connections.

The solar system was refurbished in 1974 as an ERDA project (17). The
principal changes are that the self-draining system has been replaced by an
ethylene glycol and water heat exchanger, pump and piping system to eliminate
the problems of freezing, and that five small water-to-air packaged heat
pumps have been added to extract energy from the circulating water in the
building and deliver heated air to five groups of rooms. The main objective
of the project is to develop generalised design data on solar energy assisted
heat pump systems for architects and consulting engineers.

The Henry Mathew House, Coos Bay, Oregon (18,19,20)
This house was designed and build by the owner, Henry Mathew, in 1966-67 and
is the best example of an owner-built solar house of its time. It incorpor-
ates many basic design features which can be used as the starting point in
any solar housing system. It also has the classical simplicity of the
earliest solar houses with the living room and kitchen in the south side to
absorb the winter sun, but with a long overhanging roof to shade these areas
in the summer. Fig. 4.4 shows the main features of the house and its solar
heating system. The roof collector, which is 15 m high x 24.4 m long, is
described in detail in Chapter 9, and has its performance improved by the
reflector, which consists of standard household aluminium foil stuck down
with a conventional roofing compound. Water is pumped through the pipes from
the main storage tank by a 0.25 h.p. pump, controlled initially by a roof-
mounted thermostat. The pipes drain to a 170 litre surge tank and then to

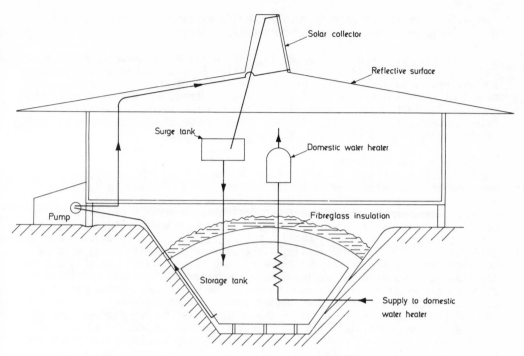

Fig. 4.4. Section Through Henry Mathew House

the storage tank whenever the pump is not working. The storage tank is
insulated from the crawl space above it, but not from the earth facing the
sides and bottom. Substantial earth heat storage - and equally heat loss -
at certain times of the year is thus made possible and autumn storage was
found to occur in the 1974-75 season. Insulated dampers in the large heat
ducts from the storage tank compartment to the living area are thermostatically controlled and can cut off all heat in the summer. There are no forced
circulation air fans. Including the 30 m^3 steel storage tank and approximately 37 m^2 of collector, the cost of the materials in 1967 was less than
$1000. Constructing the tank, his first, took Henry Mathew five weeks and
the rest of the system another eight weeks. A further 30 m^2 of free-standing
collector, erected some 20 m from the house, was added to the system in
January 1974. The atypical features of the house are as follows:-

(i)   It was built with quite standard components and not specially
       insulated although solar heating was envisaged from the start.

(ii)   Its location comparatively far north ($42\frac{1}{2}^{O}$N) in a region noted for
       overcast cloudy winter conditions.

(iii) Its combination of nearly vertical solar collector (82° to the
       horizontal) and nearly horizontal reflecting surface (8° to the
       horizontal).

(iv)   The relatively large storage capacity of the 30 m^3 storage tank.

(v) The combination of roof-mounted and free-standing collector, both with
     large reflecting surfaces immediately in front.

An unusual feature of the system is that no attempt is made during the
summer months to collect and store substantial amounts of energy for use in
the winter. Detailed results for the 1974-75 season are available (18) and
show that about 85% of the total space heating demand was met by stored solar
energy. The Mathew family responded to the environment by allowing their
interior temperature to fall below the design value of 21°C so the collector/
storage contributions diminished as winter progressed.

The Thomason Houses (21-24)
The first house designed by Thomason was a single storey house with a basement and storage space underneath the pitched roof. Erected in Washington
D.C. in 1959 it had a collector area of 78 m^2 with a total living area of
139 m^2. Thomason was one of the first designers to use the simple and comparatively inexpensive 'trickle' collector system in which water is pumped
from a storage tank to a horizontal distribution pipe at the top of the
collector. In his original system, black corrugated aluminium was used as
the absorber surface and the collector had two cover plates, one of glass and
the other a transparent polyester film. The trickle effect was obtained by
allowing the water to flow through holes in the distribution pipe directly
opposite the channels in the corrugated sheet. An open channel or gutter at
the bottom of the absorber returned the heated water to the storage tank.
The storage tank consisted of a 6.1 m^3 water tank surrounded by 50 tons of
small, 100 mm diameter, rock. The domestic water system had a 1000 litre preheater. About five days space heating storage capacity was provided by the
system and an overall performance of 95% solar heating was claimed. For
summer cooling, the water was directed over the unglazed north facing roof
channels during the night and was cooled by evaporation, convection and
radiation.

The second house, also erected in Washington D.C. in 1961, had a collector
area of 52 m^2 and a heated living area of 63 m^2. Broadly similar in concept
to the first house, it had an enhanced heating effect through a horizontal
aluminium reflector surface of 31 m^2 extending from the base of the south
facing collector. In his third Washington house erected in 1963, the storage
tank was also used as a heated indoor swimming pool and the collector system
was moved entirely onto the roof so that direct radiation from the winter sun
could enter the living room and swimming pool windows on the south side. A
fourth house, with an inferior black asphalt shingle collector system was
built next to the third, but was never fully tested and subsequently became
a storeroom. Houses five to seven were described in 1973 (23), but only
number 6, a partially heated luxury house in Mexico City, was built.

The design for number seven included a shallow roof-pond collector with a
reflector. Each night the heated water could drain to an underfloor heat
storage area, where it could warm the floor and living space. In the
mornings, a low-powered pump sent the water up to the roof. For summer
cooling, the system could work in reverse, although the exact details were
left for further design studies in any particular application. Two further
houses were built in Prince Georges County a few kilometres from Washington
D.C. In one of the houses (24), some modifications to the storage and
collection systems tried in the earlier houses have been made. The main
modification is that the rocks which surround the horizontal cylindrical 6.1 m^3

water storage tank can also be heated in winter by an oil-fired heater
through a set of copper pipes. There are also two flue pipes from the boiler
and in winter the exhaust gases pass through a flue pipe which also goes
through the rock store.

    The Thomason Houses have been widely studied, as they provide a wealth of
practical operating experience extending for nearly twenty years, and many
new solar house designs owe something in their concept to the Thomason
systems.

Solar One - The Institute of Energy Conservation, University of Delaware,
USA (25,26)
Solar One was built in 1973 as the first house to use a combination of thermal and photovoltaic solar energy in the same collector system. The other
radical feature in the house is the extensive testing of heat of fusion
thermal storage. Air is used to transfer heat from the collectors and a heat
pump is available between 'hot' and 'cold' storage systems. The philosophy
of the approach is summarised in a progress report (26) where it is pointed
out that energy is required in different 'grades' in any one domestic housing
application - low grade thermal energy for space heating or air conditioning,
higher grade thermal energy for hot water, cooking and refrigeration, and
light and electricity. Losses occur whenever one form of energy is converted
to another and it is desirable to provide as much of these forms of energy as
possible by converting them from the solar energy directly with any secondary
conversion. As the necessary data for system optimization was not available
in the early 1970's, the house was designed to allow maximum flexibility for
experimentation. A cross-section through the house is shown in Fig. 4.5.
The main single storey living area contains the living/dining room, two bedrooms, bathrooms and kitchen area. The rear, north-facing, side contains a
garage/exhibition area. As the house is designed to obtain performance data
on each component in the system, to optimize the system and to improve the
performance of the thermal-electric flat plate collectors, no attempt has
been made to live in the house or to simulate occupancy. Details on overall
performance, which would be of interest for an optimization of the collector:
storage volume ratio and storage volume:house living area ratio, are
necessarily limited by the extensive experimentation programme.

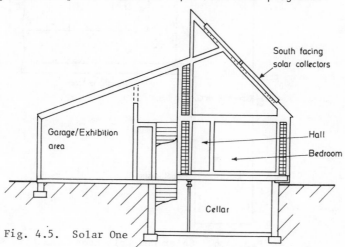

Fig. 4.5. Solar One

Twenty-four roof collectors, each nominally 1.2 x 2.43 m have been erected into the roof, which is inclined at 45° to the horizontal and faces 4.5° west of south. Three of these collectors are filled with Cadmium Sulphate/Copper Sulphide (CdS/Cu_2S) solar cells manufactured between 1968 and 1970 by the Clevite-Gould Corporation. One hundred and four of these cells were connected in series in a sub-panel and three sub-panels fit into each main collector panel. Approximately 30 W of d.c. electrical power can be produced per m^2 in full direct sunlight (about 3% efficiency). Air is circulated through the space underneath the solar cells and fins are used to improve the heat transfer characteristics to the air. Natural ventilation in the summer months is almost sufficient to keep the CdS/Cu_2S solar cells below their maximum operating temperature of 65°C. At temperatures of between 49°C and 65°C, the thermal collection efficiency lies in the 50% to 70% range with ambient temperatures between -18°C and 10°C. A cross-section through a collector is shown in Fig. 4.6. By June 1975 a total of 16 different types of collector had been tested. All had similar glazing and collector casings, but various types of selective surface, fin spacing and geometry were investigated. As a further step in the simulation, a slaved power supply was used to produce the equivalent output from the whole thermal-electric roof, which has an effective area of 57.6 m^2. Six vertical south facing thermal air collectors, each nominally 1.2 x 1.83 m and originally planned with a simple selectively coated black aluminium sheet as the heat absorbing surface, were also included in the experimental programme.

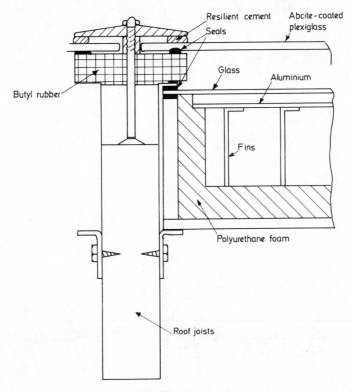

Fig. 4.6. Solar One Collector

The heat storage system occupies a relatively small volume, approximately 6.12 m^3, and consists of two outer vertical stacks of ABS pans containing sodium thiosulphate pentahydrate (Na$_2$S$_2$O$_3$.5H$_2$O), with a change of state temperature of 49°C. The central stack contains a eutectic salt, mainly sodium sulphate decahydrate (Na$_2$SO$_4$.10H$_2$O), with a change of state temperature of 12.8^{o}C, packed in tubes, each 31.75 mm in diameter and 1.83 m long. The outer system is the 'hot' store while the central stack is the 'cool' store. Both these systems have been cycled extensively and quite independently from the solar heating system. The capacity of these stores is sufficient to carry about 3 days' winter heat or 1 day's cooling in summer.

The Copper Development Association Decade 80 House, Tucson (27)

This house was designed and constructed in 1975 as an industrial market development project to create new applications for copper, brass and bronze. The CDA claim that this is the first 'real' home, as opposed to purely experimental structures, to make so much use of solar radiation, as 100% of the heating and up to 75% of the cooling load are predicted. Although it has many unusual features, such as the CDA electric vehicle in the garage which is recharged nightly, the house was built to demonstrate that all the essential components and materials needed to build an almost totally energy self-sufficient home are already available at competitive prices. A further feature is that such a house could be built by any competent local building contractor. After one year of full experimental trials, it is intended that the home will be sold for normal occupancy. Figure 4.7 shows the house with its integrated, double glazed, copper solar collector roof. The basic collector panel consists of 1.2 x 2.44 m copper sheets laminated to plywood combined with rectangular copper tubes for water circulation to the 11.4 m insulated storage tank. No detailed economic analysis has yet been published, but the CDA claimed that the solar roofing system would be paid for in about 10 years by savings in fuel.

Cooling is provided by two standard lithium bromide-water absorption units modified to use the solar-heated hot water as the heat source. This type of absorption air-conditioning unit has been a technical reality for some years, but it is only recently that it has become economically viable for residential installations, always provided that a long, trouble-free life can be obtained from the system. Silicon photovoltaic cells are also incorporated on the roof for various minor power applications, such as the low voltage supply for a small TV set and a kitchen clock. The solar cells also provide stand-by power for the home's overall security system in case of electric failure.

The roof of the connecting guest wing, which is inclined at about 40° to the horizontal, provides solar heating for the swimming pool in the spring and autumn. In the summer it is used as a simple cooling system, as the pool water can be circulated through it at night, reradiating to the cool night sky, thus maintaining the pool at comfortable temperatures in daytime. The main house roof is inclined at 27° to the horizontal to optimise on summer heat collection to provide the relatively large amount of energy needed to power the absorption cooling system. Further protection against unwanted heat gain in the summer is provided by special solar bronze tinted double glazed windows on the side of the house facing the swimming pool.

Applications in the United Kingdom

The Curtis House (28,29)
The first solar house in the UK, illustrated in Fig. 4.8, was designed by the architect-owner, Edward J.W. Curtis, who can probably claim to have lived in his own experimental solar house longer than anyone else since it was built in 1956 at Rickmansworth, near London. The house was the result of a study which he had carried out during the previous years into domestic design and environmental control systems from simple heating devices to total air conditioning. In the design stage a fundamental principle was that control of the internal environment should be obtained by a combination of solar energy and a heat pump system to provide heat and cooling, together with hot water supply and refrigeration. Curtis and others who have attempted to apply solar principles to space heating in the UK have appreciated that without a

Fig. 4.7.   CDA Decade 80 House, Tucson

Fig. 4.8. The Curtis House

very large, heavily insulated heat storage system it is only possible to
obtain a certain percentage of the total space heating requirements from
solar energy. The overall design aim was to provide a comfortable and
flexible internal requirement to offset the external temperature conditions
of a year round basis. This requirement was also linked to the aesthetic
interpretation of interior spaces where it was desired to give the maximum
feeling of spaciousness with a ground plan of only 11.2 m x 6.1 m.

The site, which is on the top of high ground overlooking a valley, was
selected to give the required orientation and to provide an unobstructed sun
track. The main spaces were planned on the south and west sides while the
east side took the entrance area, landing and two bedrooms. The construction
principle adopted was that of two free standing side walls of brick cavity
construction ends and one large panel in-filled with purpose-made timber
window walling. It was decided to provide maximum glass area on the south
front to exploit the solar gain and with one small exception the entire south
elevation consists of Plyglass clear cavity double glazing units set in
timber frames. Panels to west and north are also double glazed. Year round
air conditioning is provided by a heat pump which originally used air but was
subsequently modified to use water as the low temperature heat source.
During the first year of operation, Curtis comments that sun periods were
plotted and that it was interesting to note that during November and January
there were considerable periods of direct solar gain which helped the overall
heat build-up within the house and tended to improve the performance
generally during that time. The air flow distribution system was modified
during the next two years and the heating of ground floor areas was adjusted
for maximum efficiency with rather larger quantities of air being distributed
compared with the upper level. Average temperature levels were between
20.6°C for daytime temperatures and 22.0°C for evening periods. The build-up
of heat was found to be very rapid and except for very cold weather the heat
pump unit was switched off at approximately 23.00 hr. The high insulation
factor enabled the heat content to be retained until approximately 05.00 hr.
when the unit commenced operation giving a comfortable temperature at ground
level in the region of 19°C by 07.00 hr. The main lessons to be drawn from
this particular work are that the use of large glazed areas to obtain maximum
solar gains within the interior of a house in the UK can substantially reduce
the load on heating appliances, including a heat pump system, but on the
other hand these large glazed areas constitute a high heat loss on dull cold
days or evening periods and during winter nights. Some means of controlling
the glazed areas must be installed and used efficiently to conserve solar
build-up - even to the point of reducing daylight penetration to the absolute
minimum. In the Curtis house it is possible to have four-fifths of the
glazing covered by heavy curtaining with the remaining one-fifth giving
sufficient daylight for the interior. The total heating/cooling/solar system
has been operating satisfactorily since 1956, with overall annual running
costs approximately one-third of those of similar houses in the same neigh-
bourhood.

St. George's School, Wallasey (30,31,32)
Probably the best-known solar building in Europe, the annexe to St. George's
School, Wallasey, was designed by the late A.E. Morgan and built in 1962. It
contains a large solar wall and its ability to maintain good thermal conditions during the winter months without any conventional form of central
heating has attracted considerable attention. The Department of the
Environment have sponsored studies on the use and thermal response of the

annexe which have been carried out under the direction of Dr. M.G. Davies of
the University of Liverpool.

The main solar wall occupies the entire south facing wall of the building
and is 70 m long x 8.2 m high. A mean U-value of 3.1 W/m^2K was assumed for
all heat balance calculations. Most of the wall is double glazed with 600 mm
separation between the leaves. Each classroom, however, is provided with two
or three opening windows, which constitutes areas of single glazing. The
response time of a room in the building is about 6 days for a zero air change,
falling to 3 days with 2 air changes per hr. The depth of the building from
south to north is approximately 11.5 m. The ground floor consists of 100 mm
of screed upon 150 mm of concrete. The intermediate floor is of concrete
approximately 230 mm thick and the roof consists of approximately 180 mm concrete slab with a layer of 126 mm of expanded polystyrene above it, suitably
protected. The partition walls are of 230 mm plastered work. On the north
side at first floor level the external walls are also of 230 mm brick with
the 125 mm polystyrene external cladding. The mean U-value here is
0.24 W/m^2K. At ground floor level the external wall is partly ranch walling
and partly solar wall similar to the south side. Overall, the U-value for
the building is 1.1 W/m^2K. The only sources of heat normally operating in a
typical classroom in the annexe are due to the occupants, the electric
lighting and to solar radiation.

The older part of the school houses a similar number of students, about
300, and the two parts serve similar functions so that one acts as a control
on the other. Both staff and students in winter preferred the thermal
environment of the annexe to that of the main school, which was often
bitterly cold. However, the low ventilation rate which was necessary to
maintain the temperature in the annexe sometimes allowed the atmosphere to
become rather unpleasant. A further minor problem was noise in the summer.
To cool the building, ideally all the windows should be opened, but the staff
had found that they preferred to have high temperatures in the classroom
where they could carry on with their teaching rather than have the high noise
levels.

At the time when the building was erected the authorities insisted that a
conventional central heating system should be installed in case the solar
wall did not work. It was not until towards the end of the winter in 1973
that the central heating system was needed for the first time, but only
because vandals had broken some of the windows in the solar wall. Based on
unit costs for heating other schools in the same district, it is estimated
that about 30% of the total heating requirements of the annexe are provided
by solar energy.

The Milton Keynes House (33,34)
In 1973 the Department of the Environment gave a grant for the design of an
experimental solar heating installation in the new town of Milton Keynes,
under the direction of S.V. Szokolay, formerly of the Department of
Architecture at the Polytechnic of Central London. The aim of the project
was to test and prove the feasibility of solar heating in the UK. The
standard terrace house of the Milton Keynes Development Corporation, of which
hundreds are being built, is a thermally rather inefficient building with
reasonable insulation, but virtually no thermal inertia. It is felt that
future solar developments in the UK would use a more massive construction
and a far better thermal insulation, but with these advantages a performance

comparison with an identical house without a solar installation could not be
carried out. The solar house, illustrated in Fig. 4.9, which was taken just
prior to first occupation in March 1975, is fully instrumented for continuous
monitoring.

A feature of the design work was the extensive use made of the computer in
various parts of the process to simulate the hourly transmission of heat for
every day of the year. Although computer simulations were well known in many
applications in the United States, it is believed that this is the earliest
practical example in the United Kingdom. In this way passage of energy from
the collector to the tank and from the tank to the space heating could be
calculated together with the contribution from the auxiliary sources as
necessary. When each hourly cycle had been completed the programme could
then move to the next hour. As the programme was developed, more complicated
refinements were added, such as the stratification of temperatures in the
storage tank. It was predicted that in the period from April to September
the entire demand for space heating would be satisfied, and only in December
and January would the percentage of demand satisfied fall below 30%. For
domestic water heating, values from April to September were between 70% and
85% of the demand satisfied, although naturally there was a considerable fall
off in the winter months. The very long hot summer of 1975 caused overheating problems in the bedrooms, and it was first thought that this was
cuased by the temperature in the storage tanks immediately adjacent to the
bedrooms reaching 70°C. However, it was subsequently indicated that similar
overheating problems occurred in other adjacent non-solar heated houses, so
this is probably a basic design feature of the whole terrace. Full details
of the final design give details of the 30° sloping roof (computer simulation
called for 34°) with the 37 m^2 of solar collector (design requirements based
on 40 m^2). The total floor area is 90 m^2. The capacity of the storage tank
is 4.5 m^3 (design called for 5.2 m^3) and it is insulated with 100 mm of glass
fibre insulation.

The Higher Bebington Houses
The announcement in February 1975 of a project to build nine solar heated
houses at Higher Bebington near Liverpool, attracted considerable interest as
the planning permission had been granted by the Wirral Urban District Council
who had been involved in the approvals for the St. George's School annexe
some fifteen years earlier. The scheme is the result of a joint approach
from the glass manufacturers, Pilkington Brothers, and the Loughborough
University of Technology and will be financed by the Department of the
Environment through a local non-profitmaking housing association.

The houses will be built of high density brick with one outer wall
covered by double-glazing. This wall will absorb solar radiation which will
be reradiated into the house. The performance of the nine solar heated
houses will be compared with five conventionally built houses on the same
site. Estimated energy savings of 30% to 60% are predicted. There will be a
controlled release of heat from the wall into the living accommodation. All
nine houses will have independent fan-assisted cooling systems to augment the
natural ventilation and reduce the effect of excessive solar gain in the
summer. They are to be highly insulated and have sealed double-glazing units
in every window. One solar house and one traditional house will also have a
flat plate solar collector for domestic hot water.

Fig. 4.9. The Milton Keynes House

The Cambridge Autarchic House Design (35,36)
The project was initiated by A. Pike of the Cambridge University Department
of Architecture in 1971 with the objective of total autarchy, or self-
sufficiency.* Starting with the assumptions that the price of oil, gas,
electricity and food may quadruple over the next ten years and that the
demand for private space could increase to one acre per family and that the
three-day working week could be the norm, he felt that these trends could
stimulate families to want to produce their own power, water and food on
site. Computer studies undertaken by Pike's Technical Research Division
indicated that such a house was theoretically possible. A wind-driven
generator was included in their model simulation as well as a solar collector
occupying the entire south facing roof area, from which water could be
returned to a 40 m^3 basement storage tank. A major design feature is a
return to the concept of the Victorian Conservatory, as approximately half
the total volume under the roof of the house consists of a glazed patio area
extending over the whole south facing elevation. During the very cold season
this can be separated from the living areas and bedrooms by insulated
shutters.

Making use of available radiation data and wind data they showed that
perhaps 25% of the total annual solar radiation reaching a roof can be
utilised for space heating internally. This figure is rather lower than
would be expected, but computer simulations only put water through the solar
collectors whenever the temperature of the water leaving the solar panel
would be greater than the temperature of the water in the storage tank. The
storage tank can also be heated from the electrically generated power from
the wind generator whenever this is not being used for other domestic power
purposes and for much of the winter simulation the greater proportion of the
space heating load was carried by a wind-powered heat pump. The work has
provided an extensive series of detailed preliminary system design simulations, but by the end of 1976 no firm contract to proceed with the scheme
had been agreed.

The Granada House, Macclesfield
In January 1976 the Granada Television Company, Manchester, presented a
series of programmes on the conversion of an old coach-house into a four-
bedroomed solar heated house. Many other energy saving ideas which feature
in the most sophisticated solar housing research investigations were also
considered, such as the recovery of waste heat from the hot water and a
ventilation heat recovery system. Various types of insulation have been
tried, including 50 mm of glass fibre blanket with 50 mm expanded polystyrene
slab, 100 mm of ordinary glass fibre or 100 mm semi-rigid mineral wool slabs
on different sides (37) faced with timber planks and weather-boarding.
Based on 1975 UK Building Regulations, the gross annual space heating
requirements for the house would be 45,230 kWh, but careful attention to
double-glazing and ventilation as well as insulation reduced this to
21,910 kWh. The way in which this demand is met is shown in Fig. 4.10, which
is the theoretical annual house energy diagram based on average weather conditions, prepared by the Electricity Council Research Centre, Capenhurst.
The shaded area represents the net supplementary space heating required,
3,680 kWh, with an internal living space temperature of 19.5^{0}C, and an

*Some previous schemes for houses which make extensive use of ambient energy
were called 'autonomous' - this is strictly incorrect, as the definition of
autonomy is the right of self-government.

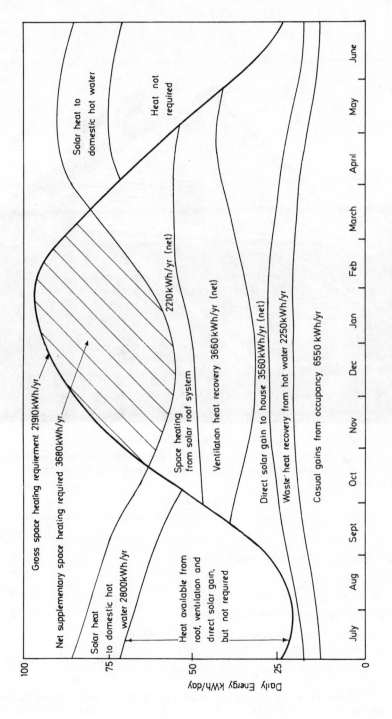

Gross space heating requirement 21910kWh/yr

Net supplementary space heating required 3680kWh/yr

Solar heat
to domestic hot
water 2800kWh/yr

Heat available from
roof, ventilation and
direct solar gain,
but not required

Space heating
from solar roof system

Ventilation heat recovery 3660kWh/yr (net)

2210kWh/yr (net)

Direct solar gain to house 3560kWh/yr (net)

Waste heat recovery from hot water 2250kWh/yr

Casual gains from occupancy 6550 kWh/yr

Solar heat to
domestic hot water

Heat not
required

Daily Energy kWh/day

100   75   50   25   0

July Aug Sept Oct Nov Dec Jan Feb March April May June

Fig. 4.10. House Energy Diagram Through the
Year Based on Average Weather
Conditions

Fig. 4.11. The Granada House

overall solar collector efficiency of 30%. The contribution from the other
heat sources to the gross space heating requirement are shown in the
following table:-

TABLE 4.1

Source	kWh/annum
Space heating from solar roof system	2210
Ventilation heat recovery	3660
Direct solar gain to house	3560
Waste heat recovery from hot water	2250
Casual gains from occupancy (cooking, lighting, etc.)	6550
TOTAL:-	18230

Fig. 4.11 shows work on the solar roof (south-west) and the north-western
side of the house with the slate-tiled outhouse, which also contains the

3000 litre solar heated water storage tank and a 2000 litre sump tank. The
north-western side of the house has only one window, while the long north-
eastern side, Fig. 4.12,has only three. Both these illustrations were taken

Fig. 4.12. The Granada House

as the solar roof was being laid. The roof, which has an area of about 45 m²,
based on the Thomason concept, is made of standard corrugated aluminium
painted with an acrylic matt black paint, single glazed with 4 mm horticultural glass. Water trickles down the channels from a horizontal perforated
pipe laid just below the ridge of the roof.

A feature of the house is the large conservatory at the ground-floor level
on the south-western side. Heated air from this conservatory can pass
directly into the top floor of the house through ducting at the bottom of the
first-floor bedroom windows. As the overall demand for space heating has
been reduced to about 20% of the gross requirement, this can be regarded as
an 80% solar heated house. Further details of a trickle system roof are
given in Chapter 9.

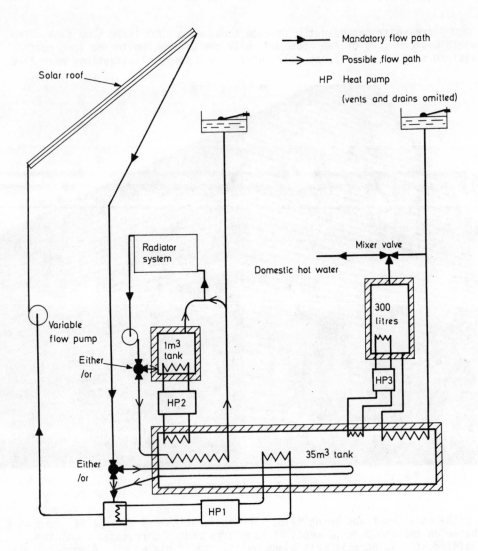

Fig. 4.13. BRE House Energy System

The Building Research Establishment Houses (38)

Three experimental houses have been proposed at the Building Research
Station, Watford, to study three major techniques of energy conservation -
solar energy, the heat pump and waste heat reclamation. In marked contrast
to the Phillips concept, which is described later, the three different sets
of options can be studied simultaneously and the BRE feel that there is no
single 'best' universal solution. The performance of the houses will be
monitored under controlled conditions with simulated occupancy. The solar
house (and the waste heat reclamation house) will be based on a timber-framed,
five person, two-storey house, the Bretton Type 47, already studied extensively by the BRE in a district heating scheme at Bretton, Peterborough. The
timber structure of these houses is prefabricated and the exterior cladding

is of brick and weatherboarding. A thickness of 92 mm of insulation in the
roof and external wall panels gives a U-value of approximately 0.29 W/m².
The roof of the solar house is pitched at 42° to the horizontal to give a
better all-round performance compared with the 22½° of the standard Bretton
Type 47. The energy system of the solar house is shown in Fig. 4.13. The
preliminary design details announced at the end of 1975 included a 22 m²
solar roof and a 35 m³ well-insulated tank located outside and below ground
level. Space heating is by radiators, but these are sized larger than normal
so that lower water temperatures can be used. Various operating modes are
selected according to the prevailing conditions. When the 35 m³ storage tank
is at a sufficiently high temperature, the radiators can draw their heat from
it. At other times they draw their heat from an insulated 1 m³ tank which,
in turn, is heated by a small off-peak electric heat pump using the 35 m³
storage tank as its low temperature source. The domestic hot water system is
fed from a 300 litre storage tank, sufficient for 24 hours of normal use.
This tank is heated either from a heat exchanger in the 35 m³ tank or through
another small off-peak heat pump. The unusual feature about the solar
collector system is that energy can be transferred to the 35 m³ storage tank
even when the temperature from the collector is lower than the tank temperature. This is achieved by the use of another heat pump.

The comparative annual energy consumptions predicted for the three houses
are shown in Table 4.2 together with figures for a standard Bretton Type 47
and a specially insulated house.

The derivation of primary energy is obtained by making conventional
assumptions on overheads and utilisation efficiencies for electricity, gas
and oil. The relatively high values of primary energy in both the Heat Pump
House and the Solar House are due to the almost exclusive use of electricity
to provide the net energy.

TABLE 4.2

Energy Consumption in GJ per annum

	Heating	Net Energy	Total Net Energy	Primary Energy
Bretton Type 47 (current building regulations)	Space Water	54.0 12.0	66.0	151.8
Bretton Type 47 (0.29 U-value)	Space Water	27.0 12.0	39.0	89.7
Heat Reclaim house	Space Water	21.0 6.5	27.5	54.4
Heat Pump house	Space Water	9.0 5.0	14.0	50.1
Solar house	Space Water	13.5	13.5	50.0

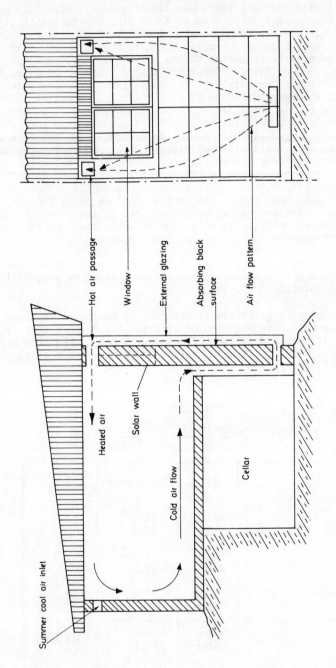

Hot air passage

Window

External glazing

Absorbing black surface

Air flow pattern

Heated air

Solar wall

Cold air flow

Cellar

Summer cool air inlet

Fig. 4.14. Trombe Three-Unit Dwelling

Applications in Other Countries

FRANCE - The Centre National de Recherche Scientifique (CNRS) (39,40,41)
The French solar housing research programme started in 1956 when the 'Trombe wall' principle was first patented. It is interesting to note the similarity between this system and that described by Professor Morse a hundred years ago (42). The basic principle is that thermally massive south facing walls, usually made of concrete, are painted black, or some other relatively heat absorbing colour such as red, dark green, or dark blue, and covered with glass on the outside, leaving an air gap between the wall and the glass. The wall is both a heat collector and a heat store. As solar radiation passes through the glass it is absorbed by the surface coating which heats the wall. As the long-wave re-radiation from the wall is trapped behind the glass, the air between the wall and the glass becomes heated. Ducts at the top and bottom of the wall allow the heated air to be fed into the room at ceiling level, while the colder air from the floor is drawn in at the bottom, as shown in Fig. 4.14. For summer cooling, the valves at the top of the wall are arranged to vent the heated air to atmosphere and the valve at the rear of the building, at ceiling level, is opened to allow cooler air to flow through. The walls are typically 300 mm to 400 mm thick. It would be possible to have other heat storage systems within the walls, such as water tanks or change-of-state chemical storage. The prototypes at Odeillo were aesthetically unattractive because they were poorly insulated and had very small south facing windows. In the latest designs the ratio of collector area to volume of the house is 0.1 m^2 per m^3 and it is now quite difficult to distinguish between the solar collectors and the windows. The latest house has been described as having the general appearance of a classical dwelling.

The French estimate that between 60 and 70% of the heating load in Mediterranean climates, such as at Odeillo, and between 35 and 50% in less favourable climates, such as Chauvency le Chateau, Meuse, can be provided by this system. The main lessons which can already be drawn are as follows:-

(i)   There are no problems of mechanical resistance to flow as with conventional roof-mounted water heaters.

(ii) There are no leakage problems.

(iii) There are no problems associated with freezing.

The latter consideration is probably most significant for the UK, but although in temperate Northern European countries a slightly greater percentage of the available solar energy lands on a vertical south facing surface in mid-winter, the scarcity of direct radiation or sunny winter days is a distinct disadvantage.

GERMANY - The Philips Minimum Energy House (43,44)
An analysis of the energy consumption in the Federal German Republic showed that about half was in the form of low temperature heat, defined as heat available at less than 100°C. The major part of this low grade heat was used in the private sector for heating buildings and providing hot water - a pattern broadly similar to that in many Northern European industrial countries. The Philips research programme concentrates on this area and has identified the measures which can be taken to reduce the consumption of conventional sources of energy into four groups as follows:-

Fig. 4.15. The Philips House

(i)   The reduction of heat losses through floors, ceilings, walls and
      windows.

(ii) The recovery of waste heat from the various domestic water systems
      and the exhaust air from the ventilation system.

(iii) The use of alternative energy sources which are not harmful to the
      environment, e.g. heat from the soil and solar energy.

(iv) The development of optimized integrated energy systems.

The aim of the research programme is to study the economic feasibility of
these resources and to develop optimized integrated energy systems.   The
experimental house, illustrated in Fig. 4.15, has been built in the grounds
of the Philips Research Laboratory in Aachen.   All the design parameters
such as its size, furnishing and household appliances were selected to match
the requirements of an average German family of four.   Two Philips P855
process computers simulate the energy demands of the family as well as controlling the various systems and processing all the data.   The various major
design features are shown in Fig. 4.16.   The emphasis is on flexibility and
many different combinations of solar heating and storage at different
temperature regimes are possible, as well as the combinations of the heat
pump system with the waste water and/or the heat from the earth.   Table 4.3
gives some basic data about the house.

The connected load of the electrical heat pump is 1.2 kW and at a temperature range between $15^{\circ}C$ and $50^{\circ}C$ its rated coefficient of performance lies

between 3.5 and 4.0.

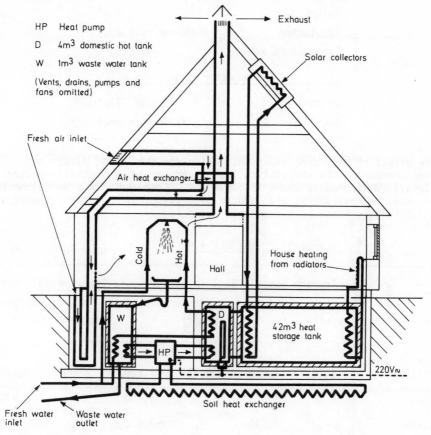

Fig. 4.16. The Philips House Energy Systems

TABLE 4.3

Some basic data

Cellar area	150 m²
Living area	116 m²
Window area	23.5 m²
Living area volume	290 m³

Long term heat storage unit

Volume	42 m³
Insulation	250 mm rock wool
Temperature range	5°C-95°C

Domestic hot water storage unit

Volume	4 m³
Insulation	250 mm rock wool
Temperature range	45°C-55°C

Waste water tank

Volume	1 m³
Insulation	100 mm rock wool

The effect of providing extra insulation for the walls, floors and ceilings, reducing the ventilation losses and providing specially coated double-glazed windows is shown in Table 4.4. Compared with a normal house, the overall thermal losses are reduced by a factor of six, and compared with a well insulated house by a factor of three.

TABLE 4.4

	Average house		Well insulated house		Experimental house	
	U_L (W/m²K)	kWh/year	U_L (W/m²K)	kWh/year	U_L (W/m²K)	kWh/year
Walls, floors, ceilings	1.23	32,630	0.48	12,600	0.14	3,630
Windows	5.80	9,970	3.3	5,700	1.5	2,570
Leakage	-	7,000	-	7,000	-	700
Controlled ventilation	-	-	-	-	-	1,400
Totals		49,600		25,300		8,300

For leakage rates it was assumed that there was one air change per hour; for controlled ventilation, one air change per hour with 80% heat recovery. The average annual energy use assumptions for a family of four were given as follows:-

(i)   Hot water, washing machine and dryer, dish washer   -   3980 kWh

(ii) Deep freeze, refrigerator   -   1095 kWh

(iii) Lighting, television and other appliances   -   1820 kWh

This gives a total of 6895 kWh, but with the waste heat recovery heat pump system, only a small percentage of the 3980 kWh used for hot water needs to be supplied as external electricity. A coefficient performance of about 3 is sufficient to save 3000 kWh, making a net demand 3895 kWh.

The energy from the earth can be used for both heating and cooling. In the heating mode, a heat exchanger consisting of a 120 m water filled plastic pipe was installed under the 150 m^2 cellar floor and by using the 1.2 kW heat pump, heat can be transferred from the soil, which has a temperature of about 7°C, to the hot water tank at a temperature of 50°C. Cooling is supplied with hardly any extra energy expenditure, as air is drawn in through a hollow cinder brick wall at cellar level. The solar collectors are integrated into the south facing roof, as shown in Fig. 4.15, and are inclined at an angle of 48° to the horizontal and cover an area of 20 m^2. Each of the 18 collector boxes contains 18 tubular evacuated glass tubes which were described in the previous chapter. The computer predictions indicated that the 10 m^2 of collector would collect between 10,000 and 12,000 kWh annually, which is more than the total heating needs of the house.

The Philips house is one of several highly instrumented experimental solar houses currently being evaluated in Europe. It will be particularly interesting to compare results from this house with the considerably less expensive but architecturally more elegant Granada House in Macclesfield.

Roof Heating Systems

For an appreciable impact to be made on the overall heating requirements of any building, a substantial area of solar collector surface is needed. Several different architectural approaches have been made to get away from the conventional flat-plate collector approach. Four systems which were under investigation in the United States in 1976 are described in the following sections.

The Stationary Reflector/Tracking Absorber System (SRTA) (45)

The basic collector, which is described in greater detail in Chapter 3, consists of a segment of a spherical mirror placed in a stationary position facing the sun. It has a linear absorber which can track the image of the sun by a simple pivoting motion about the centre of curvature of the reflector. Fig. 4.17 shows the dramatic impact that the SRTA collector could have on house design in regions where there are appreciable amounts of winter sunshine. The radical feature of the system is that water can be heated to a sufficiently high temperature for power generation and there is, therefore, the possibility of self-sufficient systems without wind-energy or direct electricity generation. Comparatively few details of any applications have been recorded but a house incorporating a SRTA collector has recently been designed and built in Colorado.

With the high proportion of diffuse radiation in the UK, together with the low levels of radiation in mid-winter, it is very unlikely that houses with SRTA system will be seen in the UK, although there would be every chance of successful operation in the Mediterranean regions.

The Roof-Pond Passive System (Hay House) (46-49)
The use of roof ponds for cooling buildings was established many years ago,
but it was not until 1967 that Hay developed a system using water enclosed in
black polyvinylchloride bags to form the roof pond. In the prototype test
building at Phoenix, the water was held in a layer about 180 mm deep which
acted both as a heat collector and storage medium. The bags were supported
on a flat metal roof, which also acted as a heat-exchanger and ceiling for
the building. At night, insulated panels were placed over the bags to prevent heat loss. During the summer, the procedure was reversed, so that
nocturnal radiation to the sky could take place, leaving a cooler pond in the
morning which was then insulated by the panels, thus providing natural
cooling to the building during the day.

The system was subsequently tested on a major two-year evaluation project
carried out by California Polytechnic State University on a 3 bedroom, 2
bathroom, single storey house at Atascadero, California. The living area of

Fig. 4.17. The SRTA System

the house, about 106 m^2, was slightly larger than the roof pond area. The movable insulation panels were electrically powered and controlled manually or by a differential temperature controller. The following extract from the report (50) on the tests indicate how successfully the building performed:-

The thermal performance of the house was very positive. The moving insulation system supplied 100% of the heating and cooling requirements of the building during the test months. During this time, the system was able to keep the indoor temperature between the extremes of 19°C and 23.3^{o}C except during special test periods or times of prototype break-down. Even during these exceptional periods, the temperature never got higher than 26°C or lower than 17°C . . .

Temperatures as high as 38°C were experienced in July and the lowest recorded temperature was -3°C in February 1974, when the monthly mean average daily outdoor temperature was 8.3^{o}C. The use of an inflatable plastic cover enabled the system to be operated in both an unglazed and single-glazed configuration. During the summer it was essential to keep the cover deflated so that the living space could be maintained at less than 27°C.

Haggard (49) believes that information from which the project can be used to investigate architectural extensions of the system which would be capable of handling other climates and spatial needs. These include multi-storey developments with movable insulation in south-facing wall channels and folding insulation panels acting as snow shields on flat roofs.

Heat Pump and Optical Systems
A limited proportion of the space heating demand can be obtained simply by installing a fan system, controlled by a differential temperature controller, in an existing, unglazed attic space. More sophisticated roof systems make use of the characteristics of conventional roof slopes by replacing a considerable area of south-facing roof by glazing, thus allowing radiation to penetrate into the attic space. Two major systems have emerged recently which use this principle. One transfers the heat in the attic directly to the rest of the house by means of a heat pump, the other uses an inexpensive reflecting optical system.

Heat Pump System (51)
In the University of Nebraska (Lincoln) and Lincoln Electric System solar house a south-facing glazed roof allows heat to be reflected in the attic, where natural circulation drives the heated air to the apex of the roof. A conventional outdoor heat pump is located near the apex and has controls which allow heat to be extracted either from the outside or from the attic. This heat is circulated to the water storage tank, which is maintained at a minimum temperature of 40°C using auxiliary heat where necessary. A vertical duct also allows direct circulation of attic air to heat the house. While the heat pump is collecting energy from the attic, the temperature there is maintained at approximately 10°C, thus high solar collection efficiencies are achieved without expensive double glazing. The stored hot water is pumped through a fan coil unit serving a conventional ducted warm air heating system. For summer cooling the heat pump is reversed, storing cold water at a minimum of 5°C. The use of conventional, commercially available systems and materials forms part of the project, which is intended to demonstrate the economic viability of the design for typical Americal Mid-West climatic conditions, where approximately 60% of the winter radiation is direct. One of

the economic aspects thrown up by the analysis is that although there is a relatively higher electrical consumption during solar collection, this is more than offset by the greatly reduced capital cost of the system, in 1975, when compared with existing flat plate collector systems. The projected design studies indicated that an overall coefficient of performance for the heat pump would be 2.72 based on a total of 800 hours operation compared with a value of 1.70 obtained from a typical conventional system installed in Lincoln.

Reflecting Optical System (52)

One system which has been described uses only plane reflecting surfaces and focusses with direct and diffuse radiation from a relatively large entrance area onto a flat plate collector typically about one fifth of the entrance area. The principles of the pyramidal optics system are shown in Fig. 4.18.

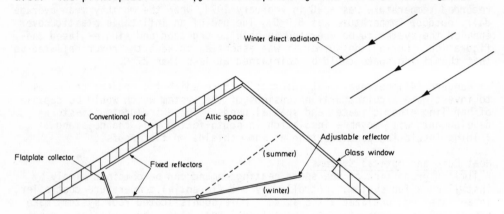

Fig. 4.18. Pyramidal Optical System

The system consists of stationary flat reflecting surfaces which make up a focussing two dimensional pyramid and a moving flat reflective surface which is adjusted for periodic variations in the sun's path - usually seasonal. A net optical gain ranging from 1.6 to 4.8 has been claimed for the system and its potential for achieving high temperatures from conventional flat plate collectors could be of considerable importance in absorption cooling applications. Various combinations of moving reflector are possible, including an externally mounted system which was used in a prototype installation at Stanford, Connecticut.

A major advantage of all glazed solar roof systems is the attractive appearance which the houses can have using conventional design and construction techniques as the attic solar collector can be taken for a normal bedroom window, as illustrated in Fig. 4.19.

Analysis of Solar Space Heating Systems

In a review (53) of Shurcliff's 'Solar Heated Buildings - A Brief Survey' (9), the comment was made that the variety of designs described will fascinate the reader and thoroughly confuse the earnest seeker after the optimum system. One of the difficulties in attempting any analysis is that

Fig. 4.19. Attic Space Heating

even with conventionally heated houses, apparently similar families in the
same district, in almost identical houses, will have wide variations in their
heating costs. An assumption, which is probably very reasonable, has to be
made that all solar houses included in the analysis are occupied by fairly
similar families who will try to get as much use of solar energy as possible
out of their particular system. A second difficulty is trying to define what
is meant by 'per cent solar heated'. Where evidence on how a family behaves
in a solar house is available, such as in the Henry Mathew house (18), it
appears that there is an acceptance of the prevailing interior temperature,
even when this falls below a level which conventional heating might have been
expected to maintain. This makes a precise assessment of how much conventional heating is actually needed rather difficult.

The main factors which could be considered in the analysis are as follows:-

  (i)   Ratio of collector area to floor area.

 (ii)   Position, angle of inclination and type of solar collector.

(iii)   Ratio of storage volume to floor area.

 (iv)   Type of storage system.

  (v)   Geographical location of building.

 (vi)   Overall insulation features, including window size, orientation and
        glazing.

(vii)   Height of the heated rooms.

Some further assumptions can now be made. The storage system can be based on
an equivalent volume of water. The overall insulation and height of the
heated rooms is hardly ever included in sufficient detail for analysis, so
variations must be neglected. This means that at any particular latitude a
series of curves could be drawn, giving the 'per cent solar heated' value
with various ratios of collector area/floor area plotted against the storage
volume/floor area ratio. This has been used as the basis of the analysis.
Fig. 4.20, showing a few curves based on latitudes less than 40°N,
illustrates the main features.

Early solar houses were generally not very well insulated and their solar
collector systems were less efficient than current designs, so the lowest
curve on Fig. 4.20, with a collector area/floor area ratio of 0.6, represents
the performances which were achieved in the 1950's. Improvements in thermal
insulation and in collector systems since then have made a considerable
difference to the performance predictions. The main tendency is that greater
'per cent solar heated' values can be obtained with relatively smaller
collectors and storage systems as shown by the the two upper curves. Considering one particular case, a performance originally represented by point
A. A slightly smaller collector area/floor area ratio would give a 100%
solar heated performance now at point B. If the same level of overall
performance were required, both the collector area and storage volume could
be halved to give point C.

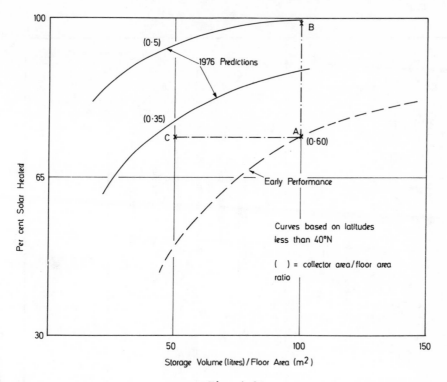

Fig. 4.20.

Figure 4.21 shows actual and predicted performance points for various
solar buildings at latitudes less than $40°N$. With storage volume (litres)/
floor area (m^2) ratios greater than 100 it can be seen that 100% solar
heating is quite possible and that very high values approaching 90% are predicted for quite low ratios of both storage volume and of collector area to
floor area.

The comparison between Fig. 4.21 and Fig. 4.22, which gives the points
for latitudes greater than $40°N$, is interesting as immediately the amount of
solar heating which can be provided is seen to be smaller. Only the Henry
Mathew house, with a significantly large volume of storage, stands out with
a relatively small collector area/floor area ratio of 0.44. In the UK, the
predicted value for the Milton Keynes house of 60% is seen to be in the right
order for its storage volume/floor area ratio. In both Fig. 4.21 and
Fig. 4.22 the predicted performances, with the improved insulation standards
and collector performance, are all following the trends illustrated in
Fig. 4.20.

It has been shown that solar space heating in buildings saves energy.
Investment in a solar energy system is always subject to interference by
Government as fuel prices are raised or lowered, but if it is considered to
be socially desirable to have buildings at least partially heated by solar
energy, it is a function of Government to see that it is economically
attractive.

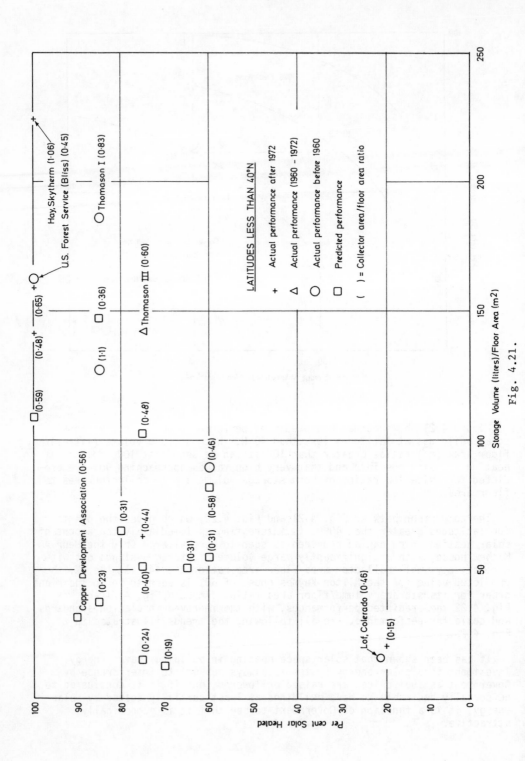

Fig. 4.21.

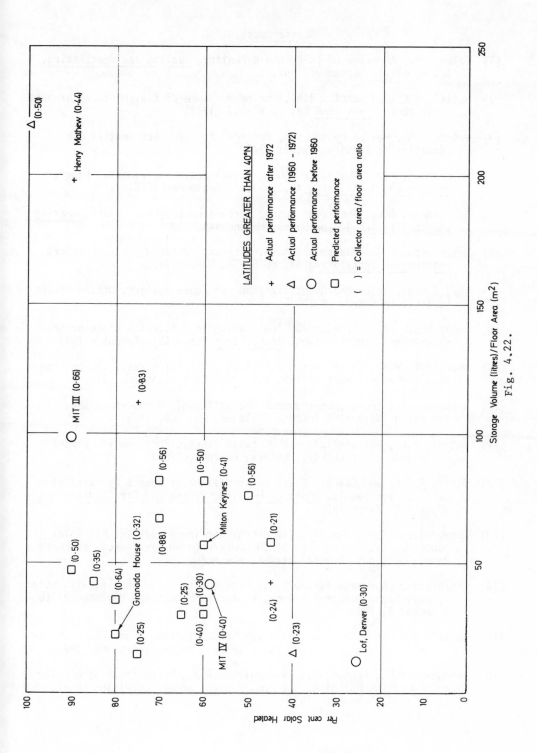

Fig. 4.22.

References

(1) Telkes, M., A review of solar house heating, Heating and Ventilating,
      46, pp 68-74, September 1949.

(2) Hottel, H.C. and Woertz, B.B., The performance of flat-plate solar-heat
      collectors, Trans ASME 64, pp 91-104 (1942).

(3) Nemethy, A., Heated by the sun, American Artisan, Residential Air
      Conditioning Section, August 1949.

(4) Hutchinson, F.W., The solar house, a full-scale experimental study,
      Heating and Ventilating 42, pp 96-7, September 1945.

(5) Hutchinson, F.W., The solar house, a research progress report, Heating
      and Ventilating 43, pp 53-55, March 1946.

(6) Hutchinson, F.W., The solar house, a second research progress report,
      Heating and Ventilating 44, pp 55-59, March 1947.

(7) Löf, G.O.G., Solar energy utilization for house heating, Office of the
      Publication Board, PB 25375, 1946.

(8) Solar house heater yields 20% fuel saving in University of Colorado
      Experimental Installation, Arch. Forum 86, p 121, February 1947.

(9) Shurcliff, W.A., Solar heated buildings - a brief survey, 19 Appleton
      St., Cambridge, Mass. 02138, USA. 8th Edition, March 1975.

(10) Steadman, P., Energy, environment and building. A report to the
      Academy of Natural Sciences, Philadelphia, CUP, 1975.

(11) Hottel, H.C., Residential uses of solar energy, Proc. World Symposium
      on Applied Solar Energy, Phoenix, Arizona, 1955.

(12) Dietz, A.G.H. and Czapek, E.L., Solar heating of houses by vertical
      wall storage panels, ASHVE J. Heating, Piping and Air Conditioning
      22, p 118, March 1950.

(13) Engebretson, C.D., Use of solar energy for space heating: MIT Solar
      House No. IV, Proceedings of UN Conference on New Sources of Energy,
      Rome, 1961, pub. United Nations, New York, 1964.

(14) Engebretson, C.D. and Ashar, N.G., Progress in space heating with solar
      energy, Paper number 60-WA-88, Winter ASME Meeting, November 27 to
      December 2, 1960.

(15) Hottel, H.C. et alia, Panel on solar house heating, Proceedings of the
      World Symposium on Applied Solar Energy, Phoenix, Arizona, 1955.

(16) Bridgers, F.H., Paxton, D.D. and Haines, R.W., Performance of a solar
      heated office building, Heating, Piping and Air Conditioning 27,
      pp 165-170, November 1957.

(17) Gilman, S.F., Evaluation of a solar energy heat pump system, ISES
      Congress, Los Angeles, Extended Abstracts, Paper 42/8, July 1975.

(18) Reynolds, J.S., Larson, M.B., Baker, M.S., Mathew, H. and Gray, R.L.,
      The Atypical Mathew solar house at Coos Bay, Oregon, ISES Congress,
      Los Angeles, Extended Abstracts, Paper 42/13, July 1975.

(19) McDaniels, D.K., Lowndes, D.H., Mathew, H., Reynolds, J.S. and
      Gray, R.L., Enhanced solar collection using reflector-solar thermal
      collector combinations, ISES Congress, Los Angeles, Extended
      Abstracts, Paper 34/11, July 1975.

(20) Mathew, H., Private communication, 1975.

(21) Thomason, H.E., Solar space heating and air conditioning in the
      Thomason house, Solar Energy 4 (4) pp 11-19 (1960).

(22) Thomason, H.E., Solar-heated house uses ¾ hp for air conditioning,
      ASHRAE J. 4 (11) pp 56-62 (1962).

(23) Thomason, H.E., Experience with solar houses, Solar Energy 10 (1)
      pp 17-22 (1966).

(24) Thomason, H.E. and Thomason, H.J.L., Solar houses - heating and cooling
      progress, Solar Energy 15 (1) pp 27-39 (1973).

(25) Böer, K.W., The solar house and its portent, Chem Tech 3, pp 394-9,
      July 1973.

(26) Böer, K.W., Higging, J.H. and O'Connor, J.K., Solar One, Two Years
      Experience, Institute of Energy Conversion, University of Delaware.
      (Presented as paper 42/3, ISES World Congress, Los Angeles 1976,
      jointly with Kuzay, T.M., Malik, M.A.S., Telkes, M. and
      Windawi, H.M.).

(27) Copper Development Association, 405 Lexington Avenue, New York,
      Press release, July 1975.

(28) Curtis, E.J.W. and Komedere, M., The heat pump, Architectural Design,
      June 1956.

(29) Curtis, E.J.W., Solar energy applications in architecture, Department
      of Environmental Design, Polytechnic of North London, February 1974.

(30) Davies, M.G., Model studies of St. George's School, Wallasey, JIHVE 39,
      p 77, July 1971.

(31) Davies, M.G., Sturrock, N.S. and Benson, A.C., Some results of measure-
      ments in St. George's School, Wallasey, JIHVE 39, p 80, July 1971.

(32) Davies, M.G., The contribution of solar gain to space heating, Sun at
      Work in Britain 3, June 1976.

(33) Szokolay, S.V., Design of an experimental solar heated house at Milton
      Keynes, UK ISES Conference on Low Temperature Collection of Solar
      Energy, April 1974.

(34) Solar heated house in Milton Keynes, Milton Keynes Development Corp.,
      Wavendon Tower, Wavendon, Milton Keynes MK17 8LX, UK.

(35) Pike, A., The Cambridge Autonomous House, UK ISES Conference on Solar
      Energy in Architecture and Planning, April 1975.

(36) Energy and Housing Symposium, Open University, Milton Keynes, Building
      Science, p 127, 31st October 1974.

(37) McLaughlin, T.P., A House For The Future, Independent Television Books
      Ltd., London, 1976.

(38) Seymour-Walker, K., Low energy experimental houses, BRE News 34,
      pp 12-13 (1975).

(39) Climatisation des Habitations Bilan Schematique des Realisations 1956-
      1972, CNRS Groupe des Laboratoires d'Odeillo (Pyrenees Orientales).

(40) Robert, J.F., Solar energy work in France, Conference on Solar Energy
      Utilization, UK ISES, July 1974.

(41) Trombe, F., Robert, J.F., Cabanat, M. and Sesolis, B., Some performance
      characteristics of the CRNS solar houses, ISES Congress, Los Angeles,
      Extended Abstracts, Paper 42/15, July 1975.

(42) Scientific American, May 13th 1882.

(43) Bruno, R., Hermann, W., Horster, H., Kersten, R. and Madhjuri, The
      utilisation of solar energy and energy conservation in the Philips
      Experimental House, ISES Congress, Los Angeles, Extended Abstracts,
      Paper 41/8, July 1975.

(44) Philips Forschungslaboratorium, Aachen, The Experimental House, 1975.

(45) Kreider, J.F., The Stationary Reflector/Tracking Absorber Solar
      Concentrator, US ISES Annual Meeting, Fort Collins, Colorado, 1974.

(46) Hay, H.R. and Yellott, J.I., A naturally air-conditioned building,
      Mechanical Engineering 92, pp 19-25, January 1970.

(47) Hay, H.R., Roof, ceiling and thermal ponds, ISES Congress, Los
      Angeles, Extended Abstracts, Paper 41/16, July 1975.

(48) Niles, P.W.B., Thermal evaluation of a house using a movable-insulation
      heating and cooling system, ISES Congress, Los Angeles, Extended
      Abstracts, Paper 42/1, July 1975.

(49) Haggard, K.L., The architecture of a passive system of diurnal heating
      and cooling, ISES Congress, Los Angeles, Extended Abstracts, Paper
      47/5, July 1975.

(50) California Polytechnic, Research evaluation of a system of natural air
      conditioning, HUD contract no. H. 2026 R., January 1975.

(51) Bourne, R.C., A volume collector-heat pump demonstration house, ISES
       Congress, Los Angeles, Extended Abstracts, Paper 14/13, July 1975.

(52) Wormser, E.M., Design, performance and architectural integration of a
       solar heating system using a reflective pyramid optical condenser,
       ISES Congress, Los Angeles, Extended Abstracts, Paper 52/8, July
       1975.

(53) ISES News No. 12, 7, June 1975.

CHAPTER 5

THERMAL POWER AND OTHER THERMAL APPLICATIONS

Solar Powered Heat Engines

The first law of thermodynamics is often expressed as follows:-

> In any enclosed system, the change of internal energy of the system is equal to the net amount of heat transferred to the system (Q) less the net external work done by the system (W). If E_2 and E_1 represent the initial and final internal energy of the system then:

$$Q - W = E_2 - E \tag{5.1}$$

To obtain a continuous work output it is necessary to bring the system back to its original state, i.e. it must pass through a cycle of operations. In equation 5.1 the net amount of heat transferred, Q, is composed of two parts. Q_1 is the heat supplied at a higher temperature than Q_2, which is the heat rejected at a lower temperature. This is a consequence of the second law of thermodynamics which states:-

> It is impossible to construct a device which will operate in a cycle and perform work while exchanging energy in the form of heat with a single reservoir.

The higher temperature reservoir is often called a source and the lower temperature reservoir a sink. In another form the second law states that heat transfer can only take place from a hotter to a cooler body.

The efficiency of the cycle is the net work output W divided by the heat input Q_1:-

$$\eta = \frac{W}{Q_1} \tag{5.2}$$

As the operation is cyclic, $W = Q_1 - Q_2$, and the efficiency can also be expressed as:-

$$\eta = \frac{Q_1 - Q_2}{Q_1} \quad \text{or} \quad 1 - \frac{Q_2}{Q_1} \tag{5.3}$$

If the absolute temperature of the source is T_1, and the sink T_2, then the cycle efficiency becomes:-

$$\eta = 1 - \frac{T_2}{T_1} \tag{5.4}$$

This is known as the ideal cycle or Carnot efficiency, after the Frenchman Sadi Carnot who was the first to develop these concepts in 1824. A more detailed treatment of this topic and its application to solar power is given by Brinkworth (1).

No engine can have an efficiency greater than the Carnot efficiency. There are various reasons for this, the main ones being losses due to friction between the moving parts and the need for a temperature difference between the source and the engine, and between the engine and the sink, so that heat transfer can take place. In practice it is very valuable to use the Carnot efficiency on a comparative basis, bearing in mind that at best the efficiency of a real engine will be about two-thirds of the Carnot efficiency.

From equation 5.4 it can be seen that the higher the source temperature T_1 becomes, the greater the efficiency for any fixed sink temperature. When this concept is considered with the characteristics of solar collectors, shown in Fig. 3.22, it can be seen that there is a conflict, as any increase in collector temperature results in a corresponding decrease in overall collector efficiency. For any given incident radiation and sink temperature it is possible to construct curves giving the ideal solar engine efficiency, which is the product of the collector overall efficiency, as shown in Fig. 3.22, and the Carnot efficiency. This has been done in Fig. 5.1 for three different collectors, taking a high incident radiation value of 900 W/m^2 and a sink temperature of 300 K (27°C).

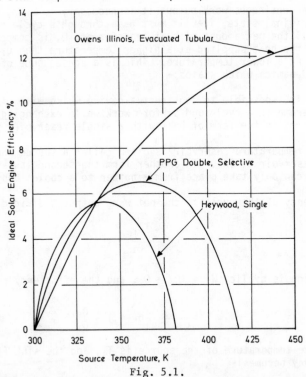

Fig. 5.1.

Up to a 35°C temperature difference between the source and the sink, the three systems have a very similar performance, indicating that for a highly inefficient operation at about 2% overall efficiency, a simple collector is quite adequate. For real overall efficiencies approaching 10% an advanced collector design is essential or some form of focussing or concentrating collector system should be provided.

Some Practical Engines

In a summary of work carried out prior to 1960, Jordan (2) commented that many ingenious solar-powered cyclic devices involving the expansion, contraction or evaporation of a solid, fluid or gas and capable of translating this effect into a periodic mechanical motion have been constructed or proposed. Most of these systems were designed for water pumping, as there is a tremendous demand for low-cost irrigation in arid regions, where there is usually a high radiation level throughout the year. Among early proposals, which are now being reconsidered as a result of the development of high efficiency collectors, was the use of a simple steam-jet injector with a flat plate collector. Steam generated through the evaporation of water was passed through a high velocity nozzle and the resultant suction near the nozzle was used for pumping water. The overall efficiency was less than 1%.

The University of Florida has been a major centre for small scale solar power generation and their work has concentrated on the development of fractional horsepower engines (3). Three main types have been studied:-

Closed cycle hot air engines in which an enclosed volume of air is displaced by a piston between hot and cold surfaces. A power piston is operated by the cyclic high pressures in the cylinder.

Open cycle hot air engines which take in atmospheric air, compress it and heat it by solar energy. The high temperature and pressure air then expands and the cycle is completed by an exhaust stroke.

Vapour engines which can operate from flat plate collectors using a conventional refrigerant, R-11 (trichloromonofluoromethane).

Both the hot air cycle engines operate with focussing collectors, which can achieve high temperatures with high efficiency, and the closed cycle engine had an output of about 250 watts with a 1.5 m diameter parabolic mirror, giving an estimated overall efficiency of just under 20%. The twin cylinder V-2 solar vapour engine (4) used three 2.8 m^2 flat plate collectors with an average collecting efficiency of over 50%. The maximum output was just under 150 watts giving an overall efficiency of about 3.5%, which is in very good agreement with predictions from the previous section.

The University of Florida has also developed a very simple solar pump in which the only moving parts are two non-return valves (5). A boiler is connected by a U-tube to a vessel containing non-return valves at inlet and outlet. The inlet valve section is connected to the water which is to be pumped. The water in the boiler is heated and turns into steam, forcing water through the outlet valve from the vessel. When the steam reaches the bottom of the U-tube, it passes rapidly into the vessel and condenses, causing the inlet valve to open as a vacuum is formed. This system is a modern version of Belidor's Solar Pump, illustrated in Fig. 1.1. Another version has been developed in England by the AERE Harwell (6) and has a very simple closed cycle hot air cylinder instead of the boiler used by the University of Florida. A variant of the basic system, the 'Fluidyne 3' pump, is shown in Fig. 5.2. One end of the U-tube is heated and the resulting change in air pressure causes the water in the output column to oscillate, forcing water through the outlet valve and drawing fresh water into the system through the inlet valve. As long as the heat is applied, the pump will continue to oscillate at its resonant natural frequency. A solar pump

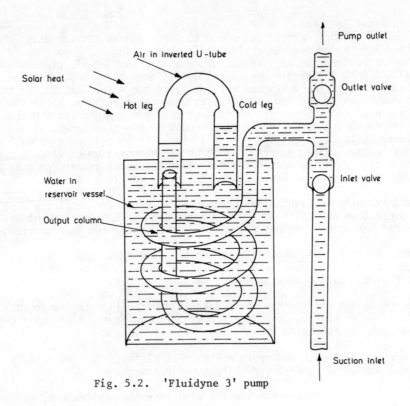

Fig. 5.2.   'Fluidyne 3' pump

developed in India (7) uses pentane vapour generated under pressure in a flat
plate solar collector as the power source. Both an air-cooled and a water-
cooled version are being studied.

Although the normal working substances in heat engines are air or steam,
certain metal alloys have the ability, when their shape is changed under
stress, to return to their original shape on being heated. A nickel-titanium
alloy, 'nitinol', has this property at about 65°C, a temperature easily
achieved by solar heating. An elegantly simple water pump based on this concept has been demonstrated in London by Frank and Ashbee (8) as a result of
earlier work on the properties of glass ceramics. The water pump itself is
no more than a ladder of inter-communicating scoops powered by the basic
engine, which is shown in Fig. 5.3. The engine rests on the rim of a vessel
containing water and consists of two vertical pillars each rigidly mounted to
a horizontal axle, with the pillars each linked to a lower horizontal rigid
bar through oppositely flexed leaf springs and oppositely bent 'nitinol'
wires. When either of the 'nitinol' wires is heated above 65°C it tends to
straighten, displacing the centre of gravity to the opposite side of the
axle and causing the device to rotate about the axle. If, as shown in
Fig. 5.3, the axle is supported on the rim of an open vessel filled with warm
water, the device oscillates about the axle, the oscillation being caused by
the 'nitinol' wires alternately straightening and displacing the centre of
gravity as they approach or dip into the water. A somewhat more elaborate
system has been developed in the United States by Banks (9).

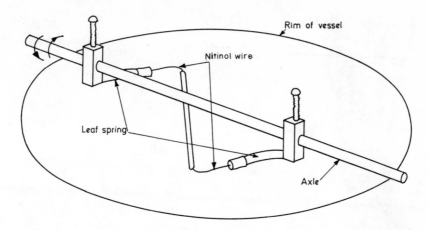

Fig. 5.3.   The Nitinol Engine

Large Scale Power Generation

Any one of the five different approaches to the large scale generation of
solar-thermal power considered in the following sections could be operating
by 1990. Several of the projects have attracted substantial research and
development efforts and some very detailed design work has taken some
sections to the prototype model stage, for example in the study of heliostat
arrays.

The Central 'Power Tower' System

A central collector system consists of a large field of steered mirrors which
reflect solar radiation to a single central receiver mounted on a large
tower. The solar radiation can be highly concentrated and high temperature
steam can be generated in the receiver. The possibility of using alternative
heat transfer fluids is also being considered. A 50 kW pilot plant has been
built at St. Illario-Nervi in Italy and can produce 150 kg/h of superheated
steam at 500°C, with a collector array consisting of 270 mirrors each 1 m in
diameter (10). More recent studies in the United States (11,12,13) have considered individual units of 100 MW capacity for intermediate and peaking
loads and the height of the tower from 300 to 450 m. A 25 MW unit is
planned in France for 1980. An impression of a field of heliostats with a
group of four power-towers, together with a central generating station is
illustrated in Fig. 5.4. Many thousands of heliostats would be required and
the spacing between the towers could be over 1 km. An economic advantage of
this system is that the individual heliostat systems could be mass produced.
The relatively small mirrors are less likely to suffer damage in high winds.
The estimated cost in 1975 for a 300 MW plant consisting of three linked
100 MW towers was $930/kW.

The Distributed Collector System

Large numbers of individual collectors are the feature of this system, which
has been called a 'solar farm'. An extensive system of insulated pipework is
necessary to collect the energy centrally to the generating station. This
could be an appropriate application for the SRTA collector, discussed in

Fig. 5.4.

Chapter 3. An impression of the distributed collector system is illustrated
in Fig. 5.5. As an alternative to individual collectors, long parabolic
troughs could be used. Both systems could be located in desert areas, but
their use is limited to regions with large amounts of direct radiation.

Ocean Thermal Energy

The use of the temperature difference between the surface of the ocean and
the colder deep water to operate a heat engine was proposed towards the end
of the nineteenth century. The oceans are natural solar energy collectors
and require no special storage systems or manufactured collectors and,
because of their enormous size, have considerable potential to compete
economically against other methods of power generation. The earliest system
was a 22 kW power plant off the coast of Cuba developed by Claude (14) in the
late 1920's. It had an overall efficiency of less than 1% and operated with
an open Rankine cycle in which the higher temperature sea water was passed
directly to a low pressure evaporator to provide steam to power the turbine.
It was uneconomic at that time, as was a subsequent larger project by the
French some twenty years later, and further work was discontinued.

Renewed interest in the concept came in the 1960's when the possibility of
using a closed Rankine cycle was suggested (15) in the United States. This
work has formed the basis for several major large scale theoretical investigations which were summarised by McGowan (16). Five different research teams
from industrial organisations and universities have selected different
systems for their various site locations. The net power outputs range from
100 to 400 MW, with overall ocean temperature differences of $17.8^\circ C$ for the
Gulf Stream off Miami to $22.2^\circ C$ for a tropical location (within 20° of the
equator). Three different working fluids were suggested, R-12/41, propane
and ammonia. The overall net cycle efficiencies of all five proposals are
very similar, ranging from 2.1 to 2.4%. Environmental studies have also been
carried out, but these have been concerned with the effects of the environment on the power plant rather than the effect of the plant on the
surrounding environment. Further studies on this aspect of the system are
clearly needed.

The basic concept is regarded as possibly the solar energy process capable
of largest feasible utilization and that by 1986 ocean thermal energy power
plant could have reached a position of dominance in the United States (16).

Satellite Solar Power Station

The use of a satellite in orbit round the earth to produce electricity which
could be fed to microwave generators for transmission to earth was first
proposed by Glaser in 1968 (17). Since then it has attracted several
detailed feasibility studies in the United States. The basic principle is
photovoltaic solar energy conversion, after concentration, from two
symmetrically-arranged solar cell arrays. The microwave generators form an
antenna between the two arrays, which directs the microwave beam to a
receiving antenna on earth. In synchronous orbit the satellite would be
stationary with respect to any particular location on earth and full use
could be made of practically continuous solar radiation. Up to 15 times
more energy is potentially available on this basis, compared with terrestrial
applications which are limited by weather conditions and daily cycles. The
system could be designed to generate from 3 to 15 GW (18).

Fig. 5.5.

Heliohydroelectric Power Generation

The concept of heliohydroelectric power generation is to convert solar energy into electricity by first transforming it into hydraulic energy. If a closed reservoir is completely sealed off from the sea, the level of the reservoir will tend to decrease as a result of evaporation. Hydro-electric generators could be placed at the reservoir end of pipes connecting the reservoir to the sea. The fall in the level of the reservoir induces a flow of water from the sea, and the potential energy caused by the difference in water levels could be transformed into electrical power. By choosing suitable water levels and power systems, it would be possible to have a continuous process. This topic has been extensively studied in Saudi Arabia by Kettani (19,20), who has measured evaporation rates and compared these with meteorological data. The possibility of building a dam across the Gulf of Bahrain is being explored (21), using the whole sealed-off Bay to create a hydraulic head from the open sea in the Gulf.

Other Focussing Systems

Cookers

Cooking by solar energy has attracted several groups of research workers since successful cookers were reported in the 18th and 19th centuries. Solar cookers can be classified into three groups. The earliest was the 'hot-box' or simple solar oven which consisted of an insulated box with a matt black interior covered with at least one transparent cover plate. Later versions had hinged lids with reflecting surfaces. In good radiation conditions temperatures greater than 100ºC can be held for several hours. The second group uses some type of focussing system to concentrate the radiation. In the 1920's Abbot (22) used a combination of a parabolic mirror and oil as the heat transfer fluid so that cooking could be achieved well into the evenings, because the higher temperatures reached by the oil gave an increased storage capacity. An extensive series of tests in India at the National Physical Laboratory under Ghai (23) led to the design of a reflector-type direct solar cooker with a spun aluminium parabolic reflector. Full details of the manu-facturing process were also given (24). Aluminized plastic film was successfully used in several cookers developed at the University of Wisconsin (25), one of which was a portable folding type mounted on a standard umbrella frame. Spherical or cylindrical concentrators have also been used and developments of the hot box and concentrating types have been reported at the University of Florida (5).

The third group uses the flat plate collector as a solar steam cooker and consists of two main components, the collector and an insulated cooking chamber, which is essentially a steam bath in which the cooking vessel can be placed. The collector has two or three transparent cover plates and has vertically rising pipes bonded to a metal sheet. These pipes are connected directly to the cooking chamber at the top of the collector. A collector with overall dimensions of 0.8 m wide by 1.55 m long was connected to a chamber containing a single 200 mm diameter by 125 mm deep aluminium cooking vessel in Haiti (26). This collector was based on a smaller unit developed at the Brace Research Institute (27). Spun aluminium parabolic cookers are commercially available and Fig. 5.6 shows one displayed at the ISES World Congress in Los Angeles in 1975. Future development of cookers will include the use of the heat pipe to transfer the heat from the collector to a longer term storage unit, so that cooking can be carried out in the late evening or early morning.

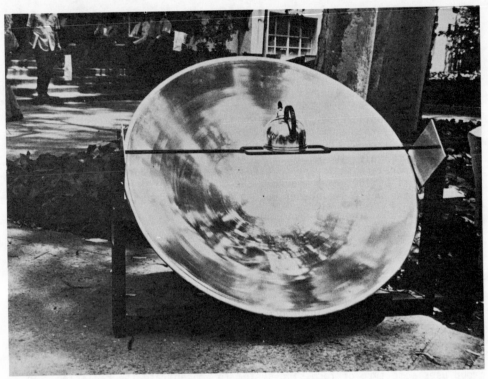

Fig. 5.6.

Furnaces

The most effective type of optical system for a solar furnace is the parabolic concentrator. There are severe practical difficulties in making a large parabolic mirror track the sun, and the alternative method is to mount the parabolic mirror in a fixed position with its axis horizontal and facing north (in the northern hemisphere). Opposite the parabola is a heliostat which tracks the sun. This method was used by Trombe (28) at the Laboratoire de l'Energie Solaire for his first major furnace, which had a mirror diameter of about 10.7 m, at Mont Louis in the Pyrenees in the 1950's. The French Centre National de la Recherche Scientifique subsequently built a 1000 kW furnace at Odeillo (29), probably the best known solar furnace in existence in the 1970's. The parabolic mirror, which is 39.6 m high and 53.3 m wide, contains 9500 individual mirrors with a total reflecting area of 1920 m^2. It faces a field of 63 heliostats, with a total mirror area of 2839 m^2. Solar furnace research is also carried out in the Soviet Union (30), the United States and Japan.

Experiments with solar furnaces have proved that it is possible to prepare refractory oxides at temperatures greater than 3000°C. Other applications include the chemical vapour deposition of materials such as molybdenum and tungsten borides (31) and high temperature phase change studies (32,33). The study of the resistance of materials to thermal shock is another application where the ability of the solar furnace to provide extremely rapid high temperatures is essential.

The philosophy behind all solar furnace research is that the majority of industrial chemical processes are based on heating by fossil fuels and it would be valuable to replace these fuels by concentrated solar radiation. However, although the Odeillo furnace has been shown to be of considerable importance as a research tool in investigations of the properties of materials at high temperatures, there are no indications that solar furnaces will ever be manufactured in quantity. Industrial applications at relatively low temperatures, such as the baking of bricks, could be of considerable interest to subtropical countries with low fossil fuel resources.

The Fresnel Lens

The concentration ratio which can be achieved by a single lens is limited, as it is difficult to manufacture accurate conventional lenses with very short focal lengths and the concentration ratio is proportional to the ratio of the diameter of the lens to its focal length. The Fresnel lens combines the advantages of a multi-lens system within a single unit as each segment is designed to concentrate the incident radiation onto a centrally positioned receiver. A further advantage is the Fresnel lens is relatively thin in a direction normal to the radiation. Figure 5.7 shows the cross-section of a

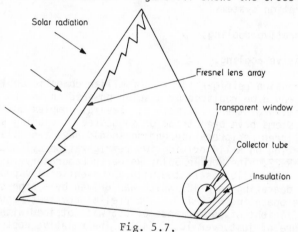

Fig. 5.7.

linear Fresnel lens, which can be mounted in a large array with one dimensional tracking (34). The performance characteristics of several linear Fresnel lens systems has been reported (35,36) and they appear to be superior to the evacuated tubular type for direct radiation up to temperatures of about 250ºC. The long term predictions for the cost of Fresnel lens systems show that they would be very competitive with oil and up to three times cheaper than electricity at 1975 prices (35). A circular Fresnel lens has been considered for low concentration in photovoltaic cells (37).

Refrigeration and Cooling

The great advantage of using solar energy in refrigeration and cooling applications is that the maximum amount of solar energy is available at the point of maximum demand. There are two quite different major applications, the first in the cooling of buildings and the second in refrigeration for food preservation.

In building applications part of the solar cooling system could be used to
provide heating outside the hot mid-summer period and the cost of the system
could be shared between the two functions. In a theoretical analysis of the
more complex systems involving combined heating and cooling for eight cities
in the United States, Lof and Tybout (38) showed that the combined system was
more economical in six of the eight cases studied. The cooling demand is at
a maximum during the early afternoon, depending on the orientation of the
building and its thermal mass, so that the storage capacity for cooling is
only a few hours in contrast to the very much greater periods required for
heating systems.

In solar heating applications the hot fluid from the collectors can often
be used to heat the interior of the building directly, but most solar cooling
applications involve a solar powered engine system. Four main methods have
been adopted as follows:

  (i)   The compression refrigeration cycle in which the refrigeration
        side is driven by a solar powered engine;

  (ii) Absorption systems;

  (iii) Evaporative cooling;

  (iv) Radiative cooling.

The basic compression refrigerator is a familiar domestic appliance and the
electric motor would be replaced by a solar powered engine for a straightforward application of the first method. Several complex compression
refrigerator systems have been tried or suggested, including a proposal (39)
that a four-cylinder reciprocating engine should have two solar-powered
cylinders operating on R-114 driving two refrigeration cylinders using
another fluid, R-22. The Mobile Solar Research Laboratory, a joint project
of the NSF and Honeywell, was fitted with a conventional vapour compression
cooling system operating on R-12, which was driven by a high-speed solar
powered turbine operating on R-113. The preliminary test results showed that
an overall coefficient of performance of 0.5 was obtained with a turbine
inlet temperature of just over $100^{\circ}C$ (40). The relative costs of the system
would decrease with size and an estimated 33-fold increase in size was
coupled with a 10-fold increase in cost. A concentrating solar collector
would considerably improve the overall performance as the turbine inlet
temperature could be increased.

The basic principles of an absorption system for cooling are shown in
Fig. 5.8. The working fluid is a solution of refrigerant and absorbent.
When solar heating is supplied to the generator some refrigerant is
vapourized and a weak mixture is left behind. The vapour is then condensed
and expands to the lower pressure evaporator, where it is vapourized and
refrigerates the external working fluid, which would be air for an air-
conditioning application. The cycle is completed in the absorber when the
refrigerant recombines with the original solution and is pumped back to the
generator. Ammonia-water systems have been successfully developed at the
University of Florida (5,41) and the University of the West Indies (42).
Because these systems provide comparatively low collector temperatures they
are more suitable for operation with currently available flat-plate
collectors. Various other developments in absorption systems were reviewed

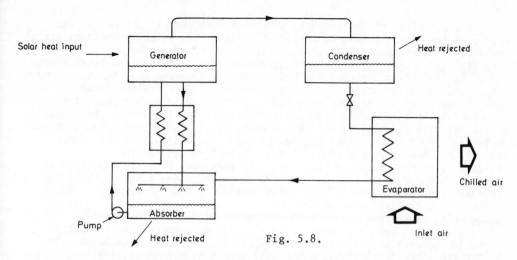

Fig. 5.8.

in 1974 and 1975 (43,44,45) including the lithium bromide-water system. A
completely different approach is the use of a desiccant wheel system with a
solar heat input to supplement a conventional gas fired burner (46). The
first installation, near Los Angeles, was completed in 1975.

Evaporative systems achieve a cooling effect through the evaporation of
water. A simple method used by Thomason (47) takes water from the house
storage tank and trickles it down the unglazed north roof. Good results have
been achieved in Australia, where water has been evaporated in the air
discharged from the buildings, and this exhaust air chills rocks in a
switched-bed rock-filled recuperator. The air flow is switched every ten
minutes and the incoming fresh air is cooled by the rocks before passing to
the building (48,49).

Radiative cooling is suitable at night with clear atmospheric conditions
and Yanigimachi (50) and Bliss (51) have used this technique by pumping water
into roof-mounted collectors. This technique has been analysed by Hay (52).
The Institute of Experimental Physics at the University of Naples has
demonstrated that radiative cooling can also take place during the day in the
presence of diffuse solar radiation (53). Selective surfaces have been
prepared with optical properties matched to the radiation emitted by the
atmosphere. This radiation has a minimum value between 8 and 13 μm, forming
an 'atmospheric window'. Theoretical predictions indicated that temperatures
about 10 to 15°C less than ambient could be achieved. The actual values
reached in a small model test were less than this, but confirmed the trend
indicated theoretically.

Applications with Heat Pumps

The principles of the heat pump were established over a hundred years ago.
By supplying energy to the heat pump, heat is transferred from a low temperature region to a higher temperature region. The earliest applications were
in refrigeration, where the food is maintained at a temperature lower than
the surroundings, while heat is rejected from the refrigerator to the
surroundings by means of an external heat exchanger.

The coefficient of performance, COP, of a heat pump is defined as the ratio of the energy output to the input. The energy output appears as useful heat at a higher temperature than the surroundings while the energy input is supplied by electricity or the direct use of fossil fuels. The total energy input to the system includes the natural energy from the environment and most heat pump installations have a COP greater than 1.0. In other words, more useful energy at the higher temperature is obtained from the system than was supplied to it through the use of electricity or fossil fuels. Theoretically it is possible to obtain a COP in the order of 20, but in practice values of between 2 and 3 are all that can be obtained (53), although a few higher values have been reported (54). For space heating applications in the British Isles heat is often required at temperatures considerably above the surroundings and it is possible to save energy by using a heat pump system instead of a conventional direct space heating system. Several experimental units have opererated over periods of many years during the past 25 years using soil, water or air as the low temperature source of heat, and a survey of this work has been published (55). The use of solar energy to augment these low temperature sources of heat appears to be attractive because the higher the input temperature to the heat pump system becomes, the smaller the conventional energy input from electricity or fossil fuels for the same net energy output. A feasibility study on a solar heated water system linked to a heat pump for the Visitor Centre at Clumber Park, Nottinghamshire (57), showed that the old underground water tanks in the greenhouses could have a new life as the heat storage system.

In the University of Nebraska-Lincoln solar house, a single-glazed south facing roof admits solar radiation directly into an attic space, instead of collecting by a conventional flat plate collector. A heat pump removes heat from this 'volume collector' and transfers the captured energy through a heat exchanger into a water storage container. Heating is then accomplished by recirculating the heated water through the house air heating system. The main advantages lie in reducing the capital cost of the collectors and in the increased efficiency of collection as there is a comparatively low temperature in the collection space. Another feature is that with a standard heat pump system there is continual cycling and a maximum energy demand when its performance is least efficient, i.e. at low ambient temperatures. By the use of a storage-coupled heat pump the unit can be smaller and can store enough energy during daylight to last through the night. The estimated COP is 2.72 which compares favourably with a value of 1.7 obtained from a typical conventional heat pump system installed in the same district (58). The economic advantages of using the heat pump in combined heating and cooling systems have also been pointed out (59).

Solar Ponds

Solar ponds have attracted considerable interest over the years in countries such as Israel (60), where there is a comparatively small seasonal variation in solar radiation. In a natural pond when solar radiation heats the water below the surface the action of convection currents causes the heated water to rise to the surface and the pond temperature normally follows the mean temperature of the surroundings. A solar pond contains concentrations of dissolved salts which gradually increase with depth, causing the density of the water to increase towards the base of the pond, which is often black. Solar radiation penetrates to the base, heating the water at this

lower level, but any convection currents are suppressed by the density
gradient. Heat losses from the surface are reduced, compared with a natural
pond, and the temperature at the bottom of the pond rises. While there are
daily fluctuations in both ambient air temperature and in the upper water
layers, the temperature at the bottom of the pond, where heat would be
extracted, remains fairly uniform (61). A solar pond is both a massive heat
collector and heat storage system and, compared with a conventional
collector and heat store, is relatively inexpensive. Tha analysis of its
performance is complex, but the general equations have been determined (62).

The range of possible applications has continued to increase, particularly
with the use of transparent membranes submerged beneath the surface of the
pond to establish a non-convective salt water layer at the top over a
separated convection zone which facilitates heat storage and extraction.
Space heating (63) and process heating (64) are two applications which show
particular promise. The concept of a shallow solar pond, about 50 mm deep,
with a transparent plastic cover, which could be emptied at night into an
underground covered storage reservoir, has been studied (65) for the production of electricity at the megawatt level. The economics of all these
applications when compared with concentrating or photovoltaic systems appear
to be competitive, although the overall efficiencies are in the order of 25%
or less.

Distillation

One of the major problems in many parts of the world is the scarcity of
fresh water and the development of inexpensive, large solar distillation
units capable of easy transportation and handling is increasingly important.
Solar distillation is another application which dates back to the 19th
century and the simplest form of water still now in use is basically
unchanged from the early designs which consisted of a shallow tray, filled
with salt or brackish water, and covered by a sloping glass cover plate. The
solar radiation heats the water in the tray and evaporates it. When the
vapour comes in contact with the colder surface of the glass it condenses,
forming fresh water which runs down the inner surface in the form of droplets
and can be collected in a trough at the lower edge. Under good radiation
conditions an output of about 4 kg/m^2 of fresh water can be obtained daily.
Two excellent summaries of solar distillation techniques were published in
1970, the first giving a comprehensive review of the history, theory,
applications and economics (66) and the second concentrating on potential
applications in developing countries (67).

One country where considerable practical experience has been gained over
many years is Australia. A guide to the design, construction and installation
of a solar still developed by the Commonwealth Scientific and Industrial
Research Organisation was published in 1965 (68) and subsequent developments
were reviewed nine years later by Cooper and Read (69). A large installation,
with an evaporating surface area of 8667 m^2 was completed on the island of
Patmos in the Aegean in 1967 (70). The average distillation rate was 3.0 kg/m^2
per day, with a maximum of 6.2 kg/m^2 at mid-summer. The first large installation designed and manufactured in the UK was a 185 m^2 unit for Aldabra in the
Indian Ocean (71) in 1970. At least one small company in the US has been

making solar stills for domestic use since the 1950's (72) and among a considerable volume of University research, the University of California,
Berkeley, has a record of over 20 years activity. This work is reviewed by
Howe and Tleimat (73), who include a design for a 37.85 m^3/day distillation
plant, with steam generated at 65.5^{o}C. Fundamental studies are also con-
tinuing in other countries such as India (74), as much of the earlier work
lacked adequate scientific measurements.

Industrial Applications

Process Heat
Any increase in the output temperature from a flat plate collector reduces
its overall efficiency and the cost effectiveness of any solar heat generating system. The potential for the large-scale use of solar energy for
industrial processes is very dependent on the operating temperature of the
process. In Australia, where there is a long tradition of solar applications
and research, a feasibility study of a typical food processing plant in
Melbourne (75) showed that it was technically practicable to phase solar
energy heating systems into existing processes. Over 50% of the annual heat
requirements could be provided by solar collectors using known techniques, as
over 70% of the requirement was at temperatures less than 100°C and there was
no significant usage above 150°C. In the food processing industry, heat is
typically generated in a central boiler house at a temperature higher than
that required for any of the processes in the plant, and then distributed as
water at 99°C or low pressure steam, 125 - 170°C, to the individual processes,
most of which operate at much lower tamperatures. The successful integration
of a solar installation with such a system requires the solar collectors to
operate at the lowest practicable temperature and the collectors, with their
associated storage if required, are coupled directly to individual processes.

Industrial solar heat generating systems must ensure that heat is available for the manufacturing process at all times and must incorporate
sufficient thermal storage and collector capacity to guarantee this under the
worst operating conditions. Alternatively, auxiliary heating must be available during periods of low radiation. Until long term summer-to-winter
storage techniques have been developed it is very unlikely that economically
viable solar heat generating systems will be developed that do not rely on
auxiliary heating. In 1975, commercially available domestic water solar
heaters operated at mean temperatures typically in the 30 - 40°C range with
an annual efficiency in the order of 40%. When these figures were applied to
a particular plant, it was found that with solar energy supplying 82% of the
annual heating load, the value of the fuel saved annually divided by the
capital cost of the system was 0.05 . An interesting figure to emerge from
this study was that if only 25% of the energy required by the entire
Australian food processing industry was supplied by the Australian solar
collector industry, this would take twenty years to achieve at 1975 production rates.

Transport
The solar-electric car of the University of Florida (5) was the first solar
powered car to operate under normal traffic conditions. The car is powered
by a 27 h.p. electric motor, which is battery-driven from NiCd and Pb acid
batteries. The batteries can be charged either by photovoltaic cells or from
a solar engine-generator system. The car has a top speed of 29 m/s on a

level road and a range of over 160 km. The long-term proposals are for the establishment of solar battery charging stations, where a discharged battery could be exchanged for a fresh one, providing an energy-free and non-polluting transportation system.

References

(1) Brinkworth, B.J., Solar Energy for Man, 115-146, Compton Press, Salisbury, Wiltshire, 1972.

(2) Zarem, A.M. and Erway, D.D. (ed.), Introduction to the Utilisation of Solar Energy, 145, McGraw Hill, 1963.

(3) Farber, E.A., The University of Florida solar energy laboratory, Conf. The Sun in the Service of Mankind, UNESCO, Paris, 1973.

(4) Farber, E.A. and Prescott, F.L., A solar powered V-2 vapor engine, Ibid.

(5) Farber, E.A., Solar energy conversion research and development at the University of Florida, Building Systems Design, February/March 1974.

(6) West, C.D., The Fluidyne heat engine, Proc. Conf. Solar Energy Utilisation, UK Section, ISES, 54-59, July 1974.

(7) Rao, D.P. and Rao, K.S., Solar water pump for lift irrigation, ISES Congress, Los Angeles, Extended Abstracts, Paper 13/12, July 1975.

(8) Frank, F.C. and Ashbee, K.H.G., Heat engine uses metal working substance, Spectrum 132, 2-4 (1975).

(9) Banks, R., Nitinol heat engines, ISES Congress, Los Angeles, Extended Abstracts, Paper 53/4, July 1975.

(10) Francia, G., Pilot plants of solar steam generation stations, Solar Energy 12 (1), 51-64 (1968).

(11) Sobin, A., Wagner, W. and Easton, C.R., Central collector solar energy receivers, Solar Energy 18 (1), 21-30 (1976).

(12) Vant-Hull, L.L. and Hildebrandt, A.F., Solar thermal power system based on optical transmission, Solar Energy 18 (1), 31-39 (1976).

(13) Blake, F.A., 100 MWe solar power plant design configuration and performance, NSF-RANN Grant No. AER-74-07570, Martin Marietta Aerospace, Denver, 1975.

(14) Claude, G., Power from the tropical sea, Mechanical Engineering 52 (12), 1039-1044, December 1930.

(15) Anderson, J.H. and Anderson, J.H. Jnr., Large-scale sea thermal power, ASME Paper No. 65-WA/SOL-6, December 1965.

(16) McGowan, J.G., Ocean thermal energy conversion - a significant solar
       resource, Solar Energy 18 (2), 81-92 (1976).

(17) Glaser, P.E., Power from the sun : its future, Science 162, 857-861,
       November 1968.

(18) Glaser, P.E., The case for solar energy, Conf. Energy and Humanity,
       Queen Mary College, London, September 1972.

(19) Kettani, M.A. and Gonsalves, L.M., Heliohydroelectric Power Generation,
       Solar Energy 14 (1972).

(20) Kettani, M.A., Climatological factors on heliohydroelectric power
       generation, Paper E38, Conf. The Sun in the Service of Mankind,
       UNESCO, Paris, 1973.

(21) Kettani, M.A., Solar energy activity in Saudi Arabia, Description of
       the Solar Energy R. & D. programs in many nations, US-ERDA, Contract
       E(04-3)-1122, February 1976.

(22) Abbot, C.G., The sun and the welfare of man, Smithsonian Institution,
       New York, 1929.

(23) Ghai, M.L., Bansal, T.D. and Kaul, B.N., Design of reflector-type
       direct solar cookers, Journal of Scientific and Industrial Research
       12A (4), 165-175 (1953).

(24) Ghai, M.L., Pandher, B.S. and Harikishandass, Manufacture of reflector-
       type direct solar cooker, Journal of Scientific and Industrial
       Research 13A (5), 212-216 (1954).

(25) Duffie, J.A., Lappala, R.P. and Lof, G.O.G., Plastics in solar stoves,
       Modern Plastics, November 1957.

(26) Alward, R., Lawand, T.A. and Hopley, P., Description of a large scale
       solar steam cooker in Haiti, Paper E46, Conf. The Sun in the Service
       of Mankind, UNESCO, Paris, 1973.

(27) Whillier, A., How to make a solar steam cooker, Brace Research
       Institute, McGill University, Do-it-yourself Leaflet L2, January
       1965.

(28) Trombe, F., Solar furnaces and their applications, Solar Energy 1, 9
       (1957).

(29) Trombe, F. and Le Chat Vinh, A., Thousand kW solar furnace built by the
       National Center of Scientific Research in Odeillo (France), Solar
       Energy 15 (2), 57-62 (1973).

(30) Arifov, U.A., Development of solar engineering in the USSR,
       Gelioteckhnika 8 (6), 3 (1972).

(31) Trombe, F., Gion, L., Royere, C. and Robert, J.F., First results
       obtained with the 1000 kW solar furnace, Solar Energy 15 (2), 63-66
       (1973).

(32) Nogucki, T., Mizuno, M. and Yamada, T., High temperature solar furnace studies, ISES Congress, Los Angeles, Extended Abstracts, Paper 23/1, July 1975.

(33) Mizuno, M., High temperature phase studies on the system Al_2O_3-Ln_2O_3 with a solar furnace, Paper 23/2, Ibid.

(34) Spitzberg, L.A. and Williams, J.K., A linear solar concentrator system, Paper 51/8, Ibid.

(35) Northrup, L.L. and O'Neill, M.J., A practical concentrating solar energy collector, Paper 51/7, Ibid.

(36) Nelson, D.T., Evans, D.L. and Bansal, R.K., Linear Fresnel Lens Concentrators, Paper 51/5, Ibid.

(37) Harmon, S., Solar optical analysis of a mass-produced plastic circular Fresnel Lens, Paper 51/11, Ibid.

(38) Lof, G.O.G. and Tybout, R.A., The design and cost of optimized systems for residential heating and cooling by solar energy, Solar Energy 16 (1), 9-18 (1974).

(39) Teagan, W.P. and Sargent, S.L., A solar-powered combined heating and cooling system, Paper EH-94, Conf. The Sun in the Service of Mankind, UNESCO, Paris, 1973.

(40) Prigmore, D., and Barber, R., Cooling with the sun's heat, Solar Energy 17 (3), 185-192 (1975).

(41) Farber, E.A., Morrison, C.A., Ingley, H.A., Clark, J.A. and Suarez, E., Solar operation of ammonia/water air conditioner, ISES Congress, Los Angeles, Extended Abstracts, Paper 44/7, July 1975.

(42) Satcunanathan, S. and Kochhar, G.S., Optimum operating conditions of ammonia-water absorption systems for flat plate solar collector temperatures, Paper 44/4, Ibid.

(43) Swartman, R.K., Vinh Ha and Newton, A.J., Review of solar-powered refrigeration, Paper 73-WA-SOL-6, ASME, 1974.

(44) Swartman, R.K., A combined solar heating/cooling system, ISES Congress, Los Angeles, Extended Abstracts, Paper 44/8, July 1975.

(45) Wilbur, P.J. and Mitchell, C.E., Solar absorption air conditioning alternatives, Solar Energy 17 (3), 193-199 (1975).

(46) Rush, W., Wurn, J., Wright, L. and Ashworth, R.A., A description of the Solar-MEC field test installation, ISES Congress, Los Angeles, Extended Abstracts, Paper 44/9, July 1975.

(47) Thomason, H.E. and Thomason, H.J.L., Solar houses/heating and cooling progress report, Solar Energy 15 (1), 27-40 (1973).

(48) Hogg, F.C., A switched bed regenerative cooling system, Proc. XIIIth
      Int. Conf. on Refrig., Washington, 4, 1971.

(49) Reed, W.R. et al, Use of RBR Systems in South Australian Schools, Aus.
      Refrig., Air Cond. and Heating 26 (12), 20-27 (1972).

(50) Yanagimachi, M., Report on two and a half year's experimental living
      in Yanagimachi Solar House II, Proc. UN Conf. on New Sources of
      Energy, Rome, 1961.

(51) Bliss, R.W. and Bliss, M.D., Performance of an experimental system
      using solar energy for heating and night radiation for cooling,
      Ibid.

(52) Hay, H.R., Roof-, ceiling- and thermal-ponds, ISES Congress, Los
      Angeles, Extended Abstracts, Paper 41/16, July 1975.

(53) Catalanotti, S., Cuono, V., Piro, G., Ruggi, D., Silvestrini, V. and
      Troise, G., The radiative cooling of selective surfaces, Solar
      Energy 17 (2), 83-90 (1975).

(54) Heap, R., Heat pumps, Ambient Energy Conference, Interbuild, London,
      1975, reported in Heating and Ventilating News, December 1975.

(55) Griffiths, M.V., Some aspects of heat pump operation in Great Britain,
      Proc. I.E.E., 104A, 262-78, 1956.

(56) Vale, B. and Vale, R., The autonomous house, 87-107, Thames and Hudson,
      London, 1975.

(57) McVeigh, J.C., Solar heating feasibility report, Appendix D, Clumber
      Park : an interpretive study, Countryside Commission, Cheltenham,
      1976.

(58) Bourne, R.C., A volume collector-heat pump demonstration house, ISES
      Congress, Los Angeles, Extended Abstracts, Paper 14/13, July 1975.

(59) Mumma, S.A. and Sepsy, C.F., A comparative experimental study of direct
      solar heating and solar assisted heat pump heating, Paper 45/2,
      Ibid.

(60) Tabor, H., Large-area solar collectors for power production, Solar
      Energy 7, 189-194 (1963).

(61) Saulnier, B., Chepurniy, N., Savage, S.B. and Lawand, T.A., Field
      testing of a solar pond, ISES Congress, Los Angeles, Extended
      Abstracts, Paper 35/1, July 1975.

(62) Weinberger, H., The physics of the solar pond, Solar Energy 8, 45-56
      (1964).

(63) Rabl, A. and Nielsen, C.E., Solar ponds for space heating, Solar Energy
      17 (1), 1-12 (1975).

(64) Styris, D.L., Zaworski, R.J., Harling, O.K. and Leshuk, J., The non-
      convecting solar pond. Some applications and stability problem
      areas. US ERDA Contract No. AT(45-1)-1830 and ISES Congress, Los
      Angeles, Extended Abstracts, Paper 35/2, July 1975.

(65) Dickenson, W.C., Clark, A.F., Day, A.J. and Wouters, L.F., The shallow
      solar pond energy conversion system, Solar Energy 18 (1), 3-10
      (1976).

(66) Talbert, S.G., Eibling, J.A. and Lof, G.O.G., Manual on solar distilla-
      tion of saline water, Office of Saline Water, US Dept. of Interior,
      Research and Development Progress Report No. 546, 1970.

(67) United Nations Dept. of Economic and Social Affairs, Solar distillation
      as a means of meeting small-scale water demands, UN Sales No. E 70
      II B1, 1970.

(68) Read, W.R.W., A solar still for water desalination, Report ED 9, CSIRO,
      Melbourne, 1965.

(69) Cooper, P.I. and Read, W.R.W., Design philosophy and operating
      experience for Australian solar stills, Solar Energy 16 (1), 1-8
      (1974).

(70) Aegean Island installs world's largest solar distillation plant,
      Civil Engineering and Public Works Review, 1005, September 1967.

(71) Porteous, A., Fresh water for Aldabra, Engineering p.490, 15 May 1970.

(72) McCracken, H., Solar stills for residential use, Paper E 6, Conf. The
      Sun in the Service of Mankind, UNESCO, Paris, 1973.

(73) Howe, E.D. and Tleimat, B.W., Twenty years of work on solar distilla-
      tion at the University of California, Solar Energy 16 (2), 97-105
      (1974).

(74) Garg, H.P. and Mann, H.S., Effect of climatic, operational and design
      parameters on the year round performance of single sloped and double
      sloped solar still under Indian arid zone conditions, ISES Congress,
      Los Angeles, Extended Abstracts, Paper 46/4, July 1975.

(75) Proctor, D. and Morse, R.N., Solar Energy for the Australian food
      processing industry, Paper 43/2, Ibid.

CHAPTER 6

METHODS OF ECONOMIC ANALYSIS

Until fairly recently the costs of most solar systems, particularly for space heating applications, have been considerably greater than those using competitive energy sources. One of the results has been that a compartively small amount of work has been carried out on the economics of solar heating and cooling systems. The classical paper, by Lof and Tybout (1), on optimizing the collector parameters to minimize the total annual heating costs for the particular geographic location, meteorological data and residential characteristics, appeared in 1970 and, in an abridged form, in 1973 (2). Their optimization technique included the effects of the number of glass covers on the collector, the collector area, the thermal storage volume and the angle of tilt of the collector. Meteorological data was based on hourly readings of solar radiation, atmospheric temperatures, cloud cover and wind velocity. This work was extended (3) in 1975 to include more recent cost equations and the effects on collector production costs of the bulk purchasing of materials and the benefits of cost reduction by performing repetitive tasks (4).

Economic studies have also been carried out to assess the impact of solar heating and cooling on a company supplying electricity to 7.5 million people in an area of about 130,000 km^2 (5). The Southern California Edison Company concluded that there were ways of combining solar heating and cooling concepts such that the use of solar energy with electrical energy was more economical, both for the Company and the customer, than the use of either alone. A regional economic study in Tennessee (6) showed that solar heating can be economical under some conditions and recommended that government incentives in the form of tax relief should be given to encourage the use of solar space heating and cooling. Basic comparisons of the effects of alternative incentives on the rate of solar energy utilization have been provided by Peterson (7), who included a consumer decision model in his analysis. The model had to meet at least three basic requirements. The first was that consumer behaviour is not based entirely on economic efficiency. The second was that there should be some allowance for differences in consumer preferences and the third that consumer behaviour may change as increased knowledge and experience is gained with solar energy.

The two aspects in the economic assessment of solar systems, the financial and the proportion of the total energy demand which can be met by solar energy, are integrated in any complete assessment. A separate analysis of various alternative approaches is given in the following sections.

Notation

C   The total capital cost of the solar heating system, including installation, ancillary equipment and collector panels, (£).

C_m The total capital cost per m^2 of collector area.

f   The annual inflation rate in the price of competitive energy,
        expressed as a fraction so that 100r is the % annual inflation
        rate.

F   Annual savings resulting from the substitution of the solar system
        for a proportion of the competitive energy system, (£).

F_c The cost of competitive energy, (£/kWh).

G_i The total annual irradiation incident on the collectors for a given
        orientation and angle of tilt (kWh/m^2).

i   The annual net effective interest rate, defined as $\frac{(1 + r)}{(1 + f)} - 1$.

        100i is the % annual net effective interest rate.

ℓ A 'loss of efficiency' factor used in Table 6.2.

M   The capital repayment factor or amortization rate, defined as Y/C with
        no inflation and Z/C with inflation. 100M is the % capital repay-
        ment factor.

n   The period of years over which the total capital cost is repaid (or
        the estimated lifetime of the solar heating system).

r   The annual interest rate on borrowed capital. 100r is the % annual
        interest rate.

T   A maintenance cost (£), used in Table 6.2.

Y   The constant annual payment necessary to repay a capital loan of £C in
        n years (£/annum).

Z   The annual payment originally considered necessary when the capital
        loan commenced. With inflation, the first payment becomes
        Z(1 + f).

η_c The efficiency of the collectors, expressed as a fraction.

η_s A system efficiency factor, which ideally is unity, but is less in
        practice as it allows for losses and differences between individual
        installations.

Standard Present Value Analysis

Conventional discounted cash flow concepts are used to establish the
present value of future savings and can be used to determine if it is
economically justifiable to invest in a solar heating system, based on the
anticipated savings in heating costs over the lifetime of the system.

The equation used to compute the constant annual payment Y necessary to
repay a capital loan C in n years at a fixed annual interest rate r is:-

$$Y = \frac{Cr(1 + r)^n}{(1 + r)^n - 1} \tag{6.1}$$

This equation is derived by considering the outstanding balance left at the end of each year when the interest and a proportion of the initial loan have been repaid. At the end of the first year the outstanding balance is:-

$$C(1 + r) - Y \tag{6.2}$$

At the end of the mth year, where m is any number between 1 and n, the outstanding balance can be expressed as:-

$$C(1 + r)^m - \frac{Y}{r} (1 + r)^m + \frac{Y}{r} \tag{6.3}$$

At the end of the nth year the outstanding balance is zero, and by substituting n for m in equation 6.3 and equating to zero the final expression becomes:-

$$C(1 + r)^n = \frac{Y}{r} (1 + r)^n - \frac{Y}{r} \tag{6.4}$$

which is the same as equation 6.1.

The Effect of Inflation

The greater the rate of inflation, the more worthwhile it becomes to make a capital investment now to reduce recurrent costs at later dates. Using the same procedure as in the previous section, with Z as the annual payment originally considered necessary and f the annual inflation rate, the outstanding balance at the end of the first year is:-

$$C(1 + r) - Z(1 + f) \tag{6.5}$$

At the end of the second year the outstanding balance is:-

$$\{C(1 + r) - Z(1 + f)\} (1 + r) - Z(1 + f)^2$$

which can be rewritten as:-

$$C(1 + r)^2 - Z(1 + f)^2 \left[1 + \frac{(1 + r)}{(1 + f)}\right] \tag{6.6}$$

and at the end of the nth year the outstanding balance is zero and the final expression becomes:-

$$C(1 + r)^n = Z(1 + f)^n \left[1 + \frac{(1 + r)}{(1 + f)} + \cdots \frac{(1 + r)^{n-1}}{(1 + f)^{n-1}}\right]$$

which can be written as:-

132   Sun power

$$C \frac{(1 + r)^n}{(1 + f)^n} = Z \left[\frac{\frac{(1 + r)^n}{(1 + f)^n} - 1}{\frac{(1 + r)}{(1 + f)} - 1} \right] \qquad (6.7)$$

The annual net effective interest rate, i, can now be introduced, where
$(1 + i) = \frac{(1 + r)}{(1 + f)}$ or $i = \frac{(1 + r)}{(1 + f)} - 1$ and equation 6.7 becomes:-

$$Z = \frac{Ci(1 + i)^n}{(1 + i)^n - 1} \qquad (6.8)$$

which has a similar form to equation 6.1

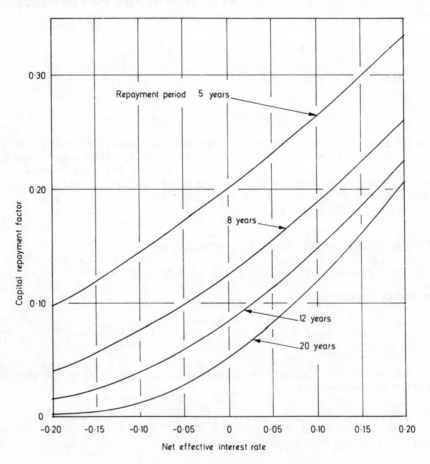

Fig. 6.1.

Over the next few decades the gradual depletion of the world's fossil fuel reserves will mean that fossil fuel costs must inevitably continue to increase. The rate of increase may be less than that experienced from 1974 to 1976, so that up to 15% annually could be reasonably anticipated. This will tend to keep the net effective interest rate fairly close to zero and will have a very considerable influence on the capital investment in a solar heating system which can be economically justified. Fig. 6.1 shows the relationship between the capital repayment factor M, defined as Z/C, and the net effective interest rate for different repayment periods. For economic viability the capital repayment factor M should be less than F/C, or the total capital cost of the system should not exceed F/M. The use of equation 6.8 or Fig. 6.1 is illustrated in the following example:-

The amount of heat which can be supplied by a solar heating system is 8000 kWh/annum. Competitive energy costs are £0.02/kWh. The anticipated increase in competitive energy cost is 14% per annum and the interest rate on a loan for the installation is 11.5% per annum. How much could be invested in the solar heating system if the loan is to be repaid in 12 years?

The effective interest rate $i \; = \; \left(\dfrac{1.115}{1.14} - 1\right) \; = \; -0.02193$

hence M $=$ 0.0719 (from equation 6.8 or Fig. 6.1).

The capital which can be invested should be less than

$\dfrac{0.02 \times 8000}{0.0719}$ $=$ £2224.

Figure 6.1 also shows that investment in solar equipment is essentially a long-term investment, as capital repayment factors smaller than 0.1 can only be achieved with a ten year period if the inflation rate is greater than the interest rate. Values for a net effective interest rate of zero, obtained when $i = r$, are sometimes known as the payback period.

Marginal Analysis

The capital repayment factor can be used to carry out a marginal analysis on any system to examine the effect of variations in collector design or system parameters, an approach which has been used by Winegarner (8). Marginal analysis can help to decide whicn collector should be used in a given system, or whether a selective surface or double glazing is worthwhile. Equation 6.8 can be expressed in terms of the total capital cost per unit area of collector, C_m, as follows:-

$$C_m \; = \; \frac{F_C \, G_i \, \eta_c \, \eta_s}{M} \tag{6.9}$$

where F_C is the cost of competitive energy, G_i the incident radiation on the collectors, η_c the collector efficiency and η_s a system efficiency factor. Differentiating with respect to collector efficiency gives:-

$$\Delta C_m \; = \; \frac{(\eta_s \, F_C \, G_i)}{M} \, \Delta \eta_c \tag{6.10}$$

The use of this equation can be illustrated by an example:-

>The application of a certain selective surface can improve the
>overall efficiency of a collector from 47% to 52%. The net
>effective interest rate is +0.03 for a period of eight years.
>The cost of competitive energy is £0.02/kWh, G_i is 930 kWh/m^2/annum
>and η_c is 0.73. What is the maximum allowable additional cost of
>the selective surface?
>
>From equation 6.8 or Fig. 6.3, M = 0.1425
>
>$$\Delta\eta_c = 52\% - 47\% = 0.05$$
>
>$$\therefore \quad \Delta C_m = \frac{0.73 \times 0.02 \times 930 \times 0.05}{0.1425} = £4.76/m^2$$

The maximum allowable additional cost for the selective surface is
£4.76/m^2 and a commercial decision can be made about the manufacturing costs.

Optimizing The Collector Area

A full treatment of this topic is beyond the scope of this book and a very
comprehensive treatment of the solar collector system and the amount of
energy which could be collected over a year has been given by Duffie and
Beckman (9). Lof and Tybout (1,2) established that two parameters had
practically no effect on the system performance, the heat transfer coefficient of the insulation on the storage tank and the heat capacity of the
collector. The optimum tilt of the collector was found to be between ten and
twenty degrees greater than latitude for sites of widely different latitude,
and variations in collector tilt from latitude to latitude plus thirty
degrees made very little difference to costs. Larger collector areas which
are needed in colder climates should be accompanied by larger storage
volumes.

The optimizing procedure commences by calculating the output from the
solar collectors, based on hourly data if possible, which is then compared
with the demand. In winter the total demand will probably not be met, while
in the summer more heat is available than is needed. The relationship
between the percentage of the total demand which can be met by solar energy
and collector area/floor area ratio follows a curve broadly similar to that
shown in Fig. 6.2. The capital costs of a solar heating system are also
shown in Fig. 6.2. There is a fixed initial sum for the pump, storage tank
and control system, while additional collectors are assumed to be at a fixed
unit cost. For each system under consideration a table can be drawn up
giving the total annual running costs, based on the discounted solar system
cost and the proportionate costs of the auxiliary system. This is used to
plot a curve relating the total cost to the collector area, from which the
optimum area can be determined.

The procedure is best illustrated by an example where the annual demand
for a floor area of 100 m^2 is assumed to be 10,000 kWh. Fig 6.2 gives the
relationship between collector area and percentage demand met. The capital
cost of the solar heating system is assumed to be £200 plus £35/m^2 of

collector area. The repayment period is twenty years at a net effective
interest rate of +5%, which gives a capital repayment factor of 0.08024. A
table can now be drawn up in which the solar cost column is obtained from
the product of the capital repayment factor and £(200 + 35A), where A is the
area of the collectors. The percentage demand met by auxiliary heating is

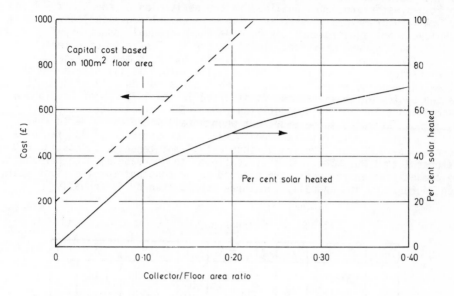

Fig. 6.2.

obtained by subtraction and, with an assumed cost of £0.025/kWh for auxiliary
heating, the total annual heating costs are obtained by adding the solar and
auxiliary costs as shown in Table 6.1.

TABLE 6.1

Collector area (m²)	Solar cost (£)	% demand met by solar	% demand met by auxiliary	Auxiliary cost (£)	Total cost (£)
10	44.13	34	66	165	209.1
20	72.22	50	50	125	197.2
30	100.3	62	38	95	195.3
40	128.4	70	30	75	203.4

The total can be plotted against collector area, as shown in Fig. 6.3, and
the optimum collector area for the system is approximately 30 m². Accurate
predictions of the optimum collector area and the effect of the other main
parameters are best obtained by a computer analysis which can use the hourly
meteorological data.

Variable Interest and Inflation Rates

Interest rates and inflation rates will not remain constant for long periods. The performance of a solar collector may deteriorate and the system may need maintenance. All these factors can be assessed by an extension of the step-by-step procedure outlined in the derivation of equations 6.1 and 6.8. A tabular method is used in which F is calculated using conventional techniques, but an allowance can be made for a gradual reduction of $\eta_c \eta_s$ by multiplying F by an 'efficiency loss' factor ℓ. The value of F tends to increase from year to year due to inflation, but this may be offset by a deterioration in collector performance. The method is illustrated in the following example, which is based on the figures for an 8.17 m^2 domestic heating system which shared the first prize in the 1975 Copper Development Association Solar Heating Competition (10). The actual total cost of the system was estimated to be £539, but when an allowance was made for savings in the original roof, this reduced to £367. The efficiency loss factor ℓ decreases by 0.02 for five years, then a full maintenance, T, costing £80 restores ℓ to 1.00. Thereafter it continues to decrease at 0.02 per annum. The competitive energy costs are assumed to be constant for the first year, then increase by 10% annually for three years, then by 5% annually.

TABLE 6.2

Year	C	r	(1 + r)C	i	F	ℓ	T
1	376.00	8	406.80	0	63.00	1.00	0
2	343.08	8	370.52	10	69.30	0.98	0
3	302.61	9	329.84	10	76.23	0.96	0
4	256.66	10	282.33	10	83.85	0.94	0
5	203.51	8	219.79	5	88.04	0.92	80
6	218.79	7	234.11	5	92.44	1.00	0
7	141.67	7	151.58	5	97.06	0.98	0
8	56.46	9	61.54	5	101.92	0.96	0

In year 3, for example, 1.10 x 69.3 = 76.23 and (302.61 x 1.09) - (76.23 x 0.96) = 256.66, the new C for year 4. A credit of £36.3 is left at the end of the eighth year.

One of the really interesting features of this detailed analysis is that during the early years of a solar installation it can show that the amount of money saved annually by the solar heating system does not have to be greater than the annual interest on the capital sum borrowed.

Any solar heating system for domestic hot water, space heating or industrial applications, will save energy. In the long term this considera- tion may outweigh all other factors.

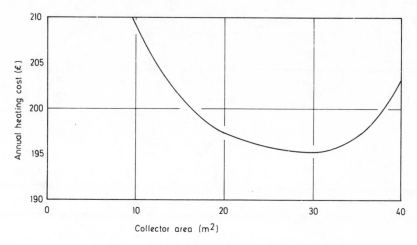

Fig. 6.3.

References

(1) Lof, G.O.G. and Tybout, R.A., Solar house heating, Natural Resources
     Journal 10, 268 (1970).

(2) Lof, G.O.G. and Tybout, R.A., Cost of house heating with solar energy,
     Solar Energy 14, 253-278 (1973).

(3) Pogany, D., Ward, D.S. and Lof, G.O.G., The economics of solar heating
     and cooling systems, ISES Congress, Los Angeles, Extended Abstracts,
     Paper 12/1 , July 1975.

(4) Behrens, H.J., The learning curve, from Cost and Optimization
     Engineering, F.C. Jelen, Ed., 1972, McGraw Hill, New York, 1972.

(5) Braun, G.W., Davis, E.S., French, R.L. and Hirshberg, A.S., Assessment
     of solar heating and cooling for an electric utility company, ISES
     Congress, Los Angeles, Extended Abstracts, Paper 12/5 , July 1975.

(6) Lunsdaine, E., Reid, R.L. and Albrecht, L., Regional economic study of
     solar heating (Tennessee), ISES Congress, Los Angeles, Extended
     Abstracts, Paper 12/2 , July 1975.

(7) Peterson, H.C., The impact of tax incentives and auxiliary fuel prices
     on the utilization rate of solar energy space conditioning, NFS-RANN
     Grant Nos. AER-09043-A01 and APR 75-18004, Utah State University,
     Logan, Utah, 1976.

(8) Winegarner, R.M., Coatings, Costs and Project Independence, Optical
     Spectra, June 1975.

(9) Duffie, J.A. and Beckman, W.A., Solar Energy Thermal Processes, John
     Wiley and Sons, New York, 1974.

(10) Copper Development Association, Orchard House, Mutton Lane, Potters
     Bar, Hertfordshire, EN6 3AP, England.

CHAPTER 7

PHOTOVOLTAIC CELLS, BIOLOGICAL CONVERSION SYSTEMS AND PHOTOCHEMISTRY

Photovoltaic Cells

The direct conversion of solar energy into electrical energy has been studied since the end of the 19th century. The early work was concerned with thermocouples of various different alloys, and efficiencies were very low, usually less than 1%. This work was reviewed by Telkes (1) in a paper written in 1953 and at that time it was felt that little more could be achieved and that efficiencies in this order were quite unsuitable for the generation of electricity. A similar pessimistic view was expressed in the United Kingdom (2). However, in 1954 the Bell Telephone Laboratories in the United States discovered that thin slices of silicon, when doped with certain traces of impurities became a factor of ten times or more efficient at the conversion of solar radiation to electricity than the traditional light sensitive materials used in earlier photocells. Since then there has been a steady history of improvement and considerably higher conversion efficiencies have been quoted - up to 16% for silicon cells and over 20% for certain new gallium arsenide cells under laboratory conditions. The use of solar cells in space applications is well known and the development for terrestrial applications is accelerating, so that by 1980 it is anticipated that solar cells will be providing a proportion of the energy requirements of many homes and buildings throughout the world.

Among the advantages listed for the modern solar cell are that is has no moving parts to wear out, has an indefinitely long life, requires little or no maintenance and is non-polluting (3). Unlike other types of electrical generator it is suitable for a wide range of power applications from less than a watt to several thousand megawatts, although a Japanese estimate considered that a 10 MW generating station made with the technology available in 1974 would require the entire world production of silicon, about 1000 tonnes (4). Silicon is widely used in the production of solar cells and is fortunately a very common material. Although solar-powered photovoltaic systems installed in Japan in 1974 were producing only an estimated 20 kW, this was considered to be more than in any other nation. The Japanese 'Project Sunshine', which started in 1974, includes plans for a 1 MW photovoltaic system by 1980, a 10 MW system by 1986 and a 100 MW system by 1991, with even larger plants operating by 2000. The National Science Foundation in the United States (5) estimated that by 1990 there should be 5000 MW at peak output manufactured annually in solar arrays and 20000 MW at peak output manufactured by 2000, which would be approximately 2% of the projected total electrical power demand.

Types of Solar Cell

The doping of a very pure semiconductor with small traces of impurities can modify its electrical properties, producing two basic types: p-type, having fixed negative and free positive charges, and n-type, having fixed positive and free negative charges. If these two types are placed together and the surface is exposed to sunlight, electrons will diffuse through the p-n junction in opposite directions giving rise to an electric current, as shown in Fig. 7.1. The earliest solar cells were made of silicon and one type of modern silicon cell is made by doping a slice cut from a single crystal of

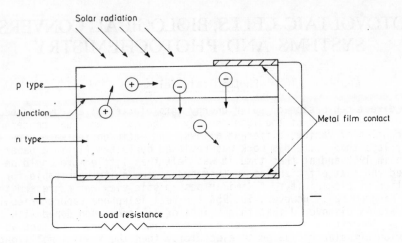

Fig. 7.1. Silicon Solar Cell

highly purified silicon with phosphorous, arsenic or antimony and diffusing boron into the upper surface, forming a 'p-on-n' cell. The front of the cell is protected by a thin glass or quartz cover. The commercial production process is complex, involving temperature control within $\pm 0.1^{\circ}C$ at $1420^{\circ}C$ during one stage, the 'pulling' of a crystal from the melt (6). Consequently silicon solar cells are expensive, costing about £20 per peak watt in 1976.

The main rival to the silicon cell for terrestrial applications is the cadmium sulphide solar cell as it is considered that the necessary processes for the mass production of cheap cadmium sulphide solar panel arrays have already been developed (7). The Institute of Energy Conversion at the University of Delaware (8) have produced Cadmium Sulphide/Cuprous Sulphide cells with conversion efficiencies close to 7% and their work indicates that this could be increased to 15%. Life expectances of over twenty years are predicted from accelerated life tests. Two types of cell with fairly similar properties are Gallium Arsenide and Indium Phosphide. Gallium Arsenide has the ability to withstand considerable concentration and up to 1000 times full sinlight has been reported by the Plessey Company in the United Kingdom and Varian Associates in California. Research is also being sponsored into organic semiconductors and Schottky barriers (metal-to-semiconductor junctions) (9).

Applications
Although the present high cost of commercially available solar cells makes them unattractive in any situation where a conventional electricity supply is available, there are already several types of application where solar cells are economically competitive. A rapidly developing area is in the provision of unmanned lights at sea. Flashing lights on buoys, lighthouses and off-shore oil rigs are increasingly being powered by solar cells, particularly in the Gulf of Mexico and the off-shore islands in Japan (10). An early application in the United Kingdom was the installation of a silicon cell

array to power a navigation light at Crossness on the River Thames in 1968 (11).   The main problem in earlier tests was the hostile marine environment and the salt in the atmosphere was found to attack some resins and plastic based mounting boards.

Automatic weather stations and other remote instruments that are difficult to reach are now being considered for solar powered operation.   Although the running costs using conventional fossil fuels may appear to be low, they have to be offset against the cost of access for maintenance and refueling.   One of the earliest uses in the United States was for powering remote radio transmitters on the top of mountains for the US Forest Service.   In Nigeria the school television programmes are intended for schools located in regions not provided with electrical power and solar cells have been used to power television receivers since 1968 (12).   The power consumption is about 32 watts from a d.c. power supply of between 30 to 36 volts.   The solar cells had an initial cost of $3100 compared with $976 for chemical batteries, but the estimated life time of the solar cells, about 10 years, would give over 25,000 hours operation, compared with the 2000 hours for the chemical batteries.   The lightweight portable power pack can also find applications, as illustrated in Fig. 7.2, which is the Ferranti array used in the recent British Everest expedition.   Battery charging is another application in the marine field for pleasure yachts and lifeboats.

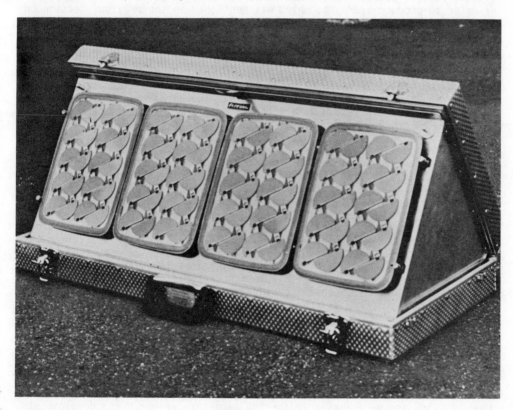

Fig. 7.2.

A commercial company in the United States, the Mitre Corporation, has installed a solar array of 1 kW peak capacity, claimed to be the largest terrestrial photovoltaic system installed by 1975 (13). Their energy storage system consisted of battery storage for short-term and peak power requirements, and an electrolysis hydrogen gas system fuel cell combination for base load and operation at night. Among other applications reported by Centralab (10) were the powering of the world's first highway callbox system in California, the use of tiny radio transmitters attached to migrating wild animals, remote snow and water gauges, and fire alarms and seismographs.

Electrical Storage

The development of storage batteries has been relatively slow and the conventional lead-acid battery is bulky and has a low power to weight ratio. The sodium-sulphur battery may be a future alternative (14). Substantial improvements in power to weight ratio could be obtained with flywheel systems. The 'super' flywheel concept using stronger and lighter materials, such as fused silica has been proposed for storing up to 70 GJ (about 20000 kWh) with a flywheel having a diameter of 5 m (15). For large scale storage another possibility is storing as high temperature heat using eutectic mixtures of metal fluorides such as NaF/MgF_2. The high temperature store could then be used as the source for a conventional thermal generating system, but the overall efficiency would be relatively low. The electrolysis of water to produce hydrogen, which was mentioned in the previous section on a small scale, could have considerable potential on a much larger scale but there are still many technical problems to be overcome (16).

Future Developments

For the silicon cell a radically different production technique is being investigated at the Tyco laboratories (17). By growing silicon crystals directly in ribbon form it is possible to eliminate the present costly process of cutting very thin wafers of silicon from large cylinders of single crystal. The process known as edge defined film-fed growth (EFG) was applied to silicon ribbon in 1971 and the first objective was to grow silicon ribbons suitable for use in solar cells. The second objective was to produce high quality silicon ribbon so that approximately 10% efficient solar cells could be achieved and the third objective was to grow the ribbon in a continuous length to indicate the basic possibility of using the technique for production of very long ribbons. These objectives have been achieved and lengths of about 2 m were approached in 1974. The final objectives are to extend these to over 30 m while retaining their efficiency at 10% and then to grow many ribbons simultaneously and continuously from a single production unit. Large scale production of these inexpensive efficient cells cannot be envisaged before 1985. One of the major objections to solar cells has been the high energy input necessary in some manufacturing processes, particularly with silicon cells. This could be overcome by using a solar furnace to manufacture silicon from silica (18).

The future development of terrestrial applications to the point where they can be used for the large scale production of electricity depends on a very considerable reduction in costs. Although a factor of over 100 sounds wildly optimistic there are other devices in common use today which cost about 100 times their present cost in the 1950's. The ballpoint pen is an excellent example and the transistor industry has also been quoted (3). As long as funds are made available for the necessary research and development work, there is no reason why the projected costs of about £2 per peak watt by 1980 and only £0.1 per peak watt by 2000 should not be achieved (3,5).

Biological Conversion Systems

Solar energy can be used by all types of plants to synthesise organic compounds from inorganic raw materials. This is the process of photosynthesis. In the process carbon dioxide from the air combines with water in the presence of a chloroplast to form carbohydrates and oxygen. This can be expressed in the following equation:-

$$CO_2 + H_2O \xrightarrow[\text{chloroplast}]{\text{sunlight}} C_x(H_2O)_y + O_2 \qquad (9.1)$$

A chlorophast contains chlorophyll, the green colouring matter of plants. The carbohydrates may be sugars such as cane or beet, $C_{12}H_{22}O_{11}$, or the more complex starches or cellulose, represented by $(C_6H_{10}O_5)_x$. All plants, animals and bacteria produce useable energy from stored carbon compounds by reversing this reaction.

Photosynthesis is an extremely important and practical method of collecting and storing solar energy and is responsible for all forms of life today. Historically the development of man can be directly traced through biological conversion systems, initially through the provision of food, then food for animals, the materials for housing and energy for cooking and heating. The commencement of industrial activities was followed by the development of agriculture and forestry to their present levels. The renewed emphasis on biological conversion systems arises from the fact that solar energy can be converted directly into a storable fuel and other methods of utilizing solar energy require a separate energy storage system. The carbohydrates can be reduced to very desirable fuels such as alcohol, hydrogen or methane, a process which can also be applied directly to organic waste materials which result from food or wood production. Compared with other methods biological conversion efficiencies are much lower, but are potentially far less expensive.

Photosynthetic Efficiency

The theoretical maximum efficiency of photosynthesis is about 27%, but under normal agricultural conditions very low efficiencies are achieved, usually less than 1%. Under very favourable conditions, conversion efficiencies of between 2 and 5% have been recorded in the field, for example a crop of bull rush millet in Australia achieved 4.2% over a growth period of 14 days in 1965 with careful control of nutrient supplies. Among other peak growth rates quoted in temperate climates are 4.3% for sugarbeet in the UK and 3.4% for maize in Kentucky, USA. Considerably lower efficiencies are achieved over longer periods of growth. Irish grasslands or forests with Sitka spruce are capable of dry matter yields greater than 16 tonnes/ha which represents an efficiency of about 0.7% (19). The Kentucky maize yield expressed as a function of the total annual radiation is only 0.8%.

The importance of improving these low efficiencies is underlined in Hall's analysis (20) for the United Kingdom, where the average energy consumption per person is equivalent to about 5 kW. With the average solar radiation level of about 110 W, 400 m^2 per person is needed at 10% solar energy conversion efficiency. The population of 55 million would therefore need only 9% of the total land area of the United Kingdom to satisfy the entire energy demand.

Energy Resources through Photosynthesis

An energy crop is grown so that its stored chemical energy can be converted
into useful energy by combustion or converted into a storable fuel. A land
crop should have as high a conversion efficiency as possible, but it does not
have to be digestible by animals or edible by humans. The entire material or
biomass of the crop can be used, including the leaves, stalks and roots. By
careful genetic selection and intensive cultivation the conversion efficiency
should reach 3% under normal conditions. An interesting development in the
United Kingdom has been the introduction of Spartina townsendii, a type of
rough grass, into intertidal mudflats. Maximum conversion efficiencies about
50% higher than other species have been reported (21).

The use of trees as energy crops has been proposed in Ireland (19) and
Australia (22). About 6% of the Irish land area consists of bogland and less
than a fifth of this area is being harvested for peat, which is either used
directly as fuel in the home or for generating electricity. In 1974 this
represented 24% of the total electrical generation. Until recently it had
been thought that bogland was unproductive, but grass, shrubs and trees have
all been successfully grown. Even with a conversion efficiency of 0.5% for
Sikta spruce, the same bogland area at present used for turf could produce
exactly half the quantity of electricity through the combustion of the trees.
An extension of the tree crop to an area about double the Irish bogland would
be sufficient to meet the entire Irish electricity demand with a continuously
renewable fuel.

An important factor in considering energy crop conversion is the energy
needed for harvesting and for fertilisers to increase the crop yields. A
detailed analysis for an isolated community of 6000 people in Australia (22)
is significant as it draws only upon existing technology and there is no
pollution because the CO_2 fixed in photosynthesis is released in the atmosphere after combustion. Two different systems were investigated, the direct
burning of chipped wood in a total energy steam raising system with a steam
turbine prime mover and the production of producer gas from chipped wood with
the burning of the gas in a total energy system with a reciprocating gas
engine. Both systems were considered to be technically feasible and the
major portion of the minerals required for forest regrowth could be recycled
by returning combustion ash to the forest area. An area of less than 6 km^2
was needed, which included the additional area for the energy required for
harvesting. The case for trees as an energy crop was put very succinctly by
T.B. Reed (23) when he stated, "I would rather go for a stroll in an acre of
forest than in an acre of solar cells".

With a suitable climate it is possible to use solar energy to dry the
energy crop. Experience with air heated solar timber kilns has been obtained
in Australia (24) with installations in Griffith and Townsville. Both
installations have prefabricated insulated kilns over a rock storage system.
The general conclusions were that solar drying took about twice as long as
conventional steam-heated kilns, but only half the time needed for air
drying. Many applications in food drying, such as fruit, vegetables and
grain have also been reported (25,26).

In oceans the production of organic matter by photosynthesis is generally
limited by the availability of nutrients and they have been compared to
deserts because of their low productivity. However, there are a few areas
where natural flows bring the nutrients from the bottom of the ocean to the

surface so that photosynthesis can take place. A preliminary study has been carried out on the cultivation of giant kelp (macrocystis pyrifera), a large brown seaweed, in an area of approximately 600,000 km^2 off the west coast of the United States. Preliminary estimates have predicted yields approaching 0.5 tonne/ha which would represent an energy production equivalent to 2% of the 1970 US production (25).

A new area in photobiological conversion research is the utilization of natural products from photosynthetic marine micro-organisms (26). Many of these organisms are capable of hydrogen photoproduction and nitrogen fixation, but the conversion efficiency is still too low. Several methods of increasing the efficiency have been suggested but a considerable amount of basic research is still needed.

A machine for converting grass and leaves into edible protein has been developed at the Rothamsted Agricultural Research Station (27). It can convert one ton of leaf into enough protein to meet the daily needs of 300 people by separating the fibre from the protein in leaves and grass. The edible protein has 6 times the protein content of an equivalent amount of steak. Another British development is the nutrient film technique (28) in which plants are grown in plastic troughs, closed at the top except where the plants emerge. The bottom of the troughs contains a thin film of water containing the nutrients. The method shows considerable possibilities, for the use of flat plate collectors for heating and wind energy systems for both heating and pumping. Among its advantages are that it eliminates the need for soil cultivation and sterilisation in greenhouses.

Full use is made of the available solar energy in a greenhouse design specially developed for colder regions at the Brace Research Institute (29). It is oriented on an east-west axis with a large transparent south-facing roof. The rear, inclined north-facing wall is insulated with a reflective cover on the inner face. Heating requirements were reduced by up to 40% compared with a standard, double layered plastic covered greenhouse and increased yields of tomato and lettuce crops were reported. A sectional view of the greenhouse is shown in Fig. 7.3.

Conversion of Solid Organic Materials to Fuels

There are several well-known processes which are ideally suitable for producing fuels from energy crops. With aerobic fermentation, materials containing simple sugars and starches can be used to produce ethyl alcohol or ethanol. The process needs considerable development before it can compete economically with conventional fuels. Anaerobic fermentation, the fermentation of organic matter in the absence of oxygen, has been used for many years for the treatment of domestic sewage, leading to the generation of significant amounts of methane. Several large sewage treatment plants have used their own methane for their energy requirements. The possibility of using other organic wastes from crop growing, animals and food processing, as well as energy crops has attracted attention. An estimate of between 10 and 20% of the gas consumption of the United States could be provided from collectible organic wastes by the anaerobic fermentation process according to a paper prepared in 1972 (32). In the pyrolysis process the organic material is heated to temperatures between 500 and 900°C at ordinary pressures in the absence of oxygen, producing methanol, which was a byproduct of charcoal in the last century. Methanol is used extensively as a fuel for high performance racing cars and is being studied as an additive at the MIT Energy Laboratory (23).

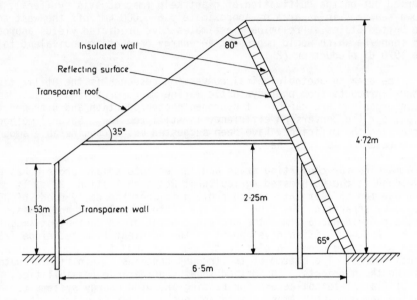

Insulated wall

Reflecting surface

Transparent roof

35°

80°

4·72m

2·25m

1·53m   Transparent wall

65°

6·5m

Fig. 7.3.

Photochemistry

The direct conversion of solar energy into stored chemical free energy has attracted research workers for many years. A review of this early work has been published by Archer (33) who also defined the fundamental photochemical work that remains to be done (34). Approximately half the total solar radiation which reaches the earth arrives as visible light and can be used in various photochemical reactions. The other half, which occurs in the infra-red region, cannot make a useful contribution as its energy concentration is too low. The maximum overall efficiency of any photochemical energy conversion is limited to about 30%, however, as a proportion of the higher energy photons of shorter wavelengths have some energy degraded as heat during the reaction. The majority of photochemical reactions are exothermic - giving out energy - and are not suitable for converting solar radiation into stored chemical energy. The known endothermic - energy storing - reactions which occur with visible light are, in theory, capable of producing valuable chemical fuels, but a major problem has been that most of these endothermic reactions reverse too quickly to store the energy of the absorbed light. Other problems include undesirable side reactions and the high cost of the relatively scarce original material. This is relatively unimportant as the original material would be regenerated when the reaction is reversed and the stored energy is released.

One particular process which has attracted attention for many years is the possible combination of carbon dioxide and water to produce various hydro-carbons such as methane. Another possibility is the photosensitised decomposition of water to hydrogen and oxygen. This has already been achieved, although with very low efficiencies, by metal cations such as cerium and europium and the use of titanium dioxide electrodes has also been reported (35). Certain organic substances can be photoreduced in water - again with very low efficiencies.

The concept of combining the photochemical and electrochemical processes in an electrical storage battery which can be recharged directly by solar radiation is very attractive. Several examples, such as the iron-thionine system (36) are well known but have efficiencies in the order of 0.1%. In this process the bulk of the solution undergoes a photochemical change and the subsequent change in the oxidation-reduction system causes the potential. An alternative approach is to coat the electrode of the half-cell with a layer of dye or an inorganic substance such as titanium dioxide. The direction of electron flow is reversed when the electrodes are irradiated.

There is considerable optimism about the possibilities of the photochemical utilisation of solar energy as there is such a wide range of options and a good theoretical background (35).

References

(1)   Telkes, M., Solar thermoelectric generators, Journal of Applied Physics 25 (6), 765-777, June 1954.

(2)   Utilization of Solar Energy, Report of the NPL Committee (UK), published in Research 5, 522-529 (1952).

(3)   Solar energy : a UK assessment, UK Section, ISES, London, 1976.

(4)   Cohen, C.L., In Japan, clean energy comes first, Electronics 47 (7), 104-5 (1974).

(5)   An Assessment of Solar Energy as a National Energy Resource, NSF/NASA Solar Energy Panel, 1972.

(6)   Currin, C.C., Ling, K.S., Ralph, E.L., Smith, W.D. and Stirn, R.J., Feasibility of low cost silicon solar cells, 9th Photovoltaic Specialists Conference, Maryland, May 1972.

(7)   Mytton, R.J., Progress in the development of cadmium sulphide terrestrial solar batteries, Proc. Conf. on Photovoltaic Cells, UK Section, ISES, November 1974.

(8)   Boer, K.W. et al, CdS/Cu_2S solar cells for large scale terrestrial applications, ISES Congress, Los Angeles, Extended Abstracts, Paper 21/7, July 1975.

(9)   Saha, H., Metal-Cu_2S Shottky barrier solar cell, Paper 21/9, Ibid.

(10) Rosenblatt, A.I., Energy crises spurs development of photovoltaic power sources, Electronics 47 (7), 106 (1974).

(11) Richards, E.R., The use of solar cells in the maritime field, Proc.
      Conf. on Photovoltaic Cells, UK Section, ISES, November 1974.

(12) Polgar, S., Use of solar generators in Africa for broadcasting equip-
      ments, ISES Congress, Los Angeles, Extended Abstracts, Paper 13/1,
      July 1975.

(13) Haas, G.M., Bloom, S. and Cherdak, A., Experience to date with the
      Mitre Terrestrial Photovoltaic Energy System, Paper 21/3, Ibid.

(14) A sort of battery, New Scientist 63 (905), 77 (1974).

(15) Post, R.F. and Post S.F., Flywheels, Scientific American 17, December
      1973.

(16) Dawson, J.K., Prospects for hydrogen as an energy resource, Nature 249,
      724-726, 21 June 1974.

(17) Mlavsky, A.I., Press release comments, Tyco Laboratories Inc., Waltham,
      Mass. 02154, 1974.

(18) Kobayashi, M., A proposal for a consistent process of manufacturing
      silicon for solar cells from silica by the use of solar energy, ISES
      Congress, Los Angeles, Extended Abstracts, Paper 21/6, July 1975.

(19) Lalor, E., Solar Energy for Ireland,National Science Council, Dublin,
      February 1975.

(20) Hall, D.O., Photobiological energy conversion, Sun at Work in Britain
      1, 14-17 (1974).

(21) Long, S., The photosynthetic potential of C_4 - plants in cool temperate
      ecosystems with particular reference to Spartina townsendii, Proc.
      Conf. Solar Energy : Biological conversion systems, UK Section, ISES
      and British Photobiological Society, June 1975.

(22) Burrow, A.C. and Taylor, L.E., Growing Kilowatts - bring back the axes,
      Ballarat Institute of Advanced Education, Australia, 1974.

(23) Reed, T.B., Bioconversion of solar energy, Testimony before the US
      House of Representatives Subcommittee on Energy, 18 June 1974.

(24) Read, W.R. and Czech, J., Operating experience with a solar timber
      kiln, ISES Congress, Los Angeles, Extended Abstracts, Paper 46/2,
      July 1975.

(25) Akyurt, M. and Selouk, M.K., A solar drier supplemented with auxiliary
      heating systems for continuous operation, Solar Energy 14, 313-320
      (1973).

(26) Peart, R.M. and Foster, G.H., Grain drying with solar energy, ISES
      Congress, Los Angeles, Extended Abstracts, Paper 46/1, July 1975.

(27) Wolf, M., Utilization of solar energy by bioconversion - an overview,
      Testimony before the US House of Representatives Science and
      Astronautics Committee, 13 June 1974.

(28)   Mitsui, A., Long range concepts : applications of photosynthetic
       hydrogen production and nitrogen fixation research, Conf. Capturing
       the Sun through Bioconversion, Washington, March 1976.

(29)   Making curry and haggis from leaves, The Times, 1 April 1976.

(30)   Cooper, A., Papers on the nutrient film technique, The Glasshouse Crops
       Research Institute, Littlehampton, Sussex, 1975-6.

(31)   Lawand, T.A., Alward, R., Saulnier, B., and Brunet, E., The development
       and testing of an environmentally designed greenhouse for colder
       regions, Solar Energy 17 (5), 307-312 (1975).

(32)   Schneider, T.R., Substitute Natural Gas from Organic Materials, ASME
       Winter Meeting, New York, 27-30 November 1972.

(33)   Archer, M.D., Photochemical Aspects of Solar Energy Conversion, in
       Photochemistry 6, ed. D. Bryce-Smith, Chemical Society, Specialist
       Periodical Report, London, 1975.

(34)   Archer, M.D., The outlook for photochemical energy conversion, ISES
       Congress, Los Angeles, Extended Abstracts, Paper 22/3, July 1975.

(35)   Porter, G., Photochemical energy conversion, Sun at Work in Britain 1,
       12-13 (1974).

(36)   Clark, W.D.K. and Echert, J.A., Photogalvanic Cells, Solar Energy 17
       (3), 147-150 (1975).

CHAPTER 8

WIND POWER

Introduction

Energy from the wind is derived from solar energy, as a small proportion of the total solar radiation reaching the earth causes movement in the atmosphere which appears as wind on the earth's surface. The wind has been used as a source of power for thousands of years, both on land and at sea. Sailing ships were first reported in ancient Egypt nearly five thousand years ago and reached their peak towards the middle of the nineteenth century with the development of the fast international trading clipper ships. However, by the beginning of the twentieth century fossil-fueled steam-engined ships had become firmly established and although the wooden sailing ships of that era could compete with steam, they continued to decline as the engine-powered steel ships improved technologically and by the nineteen-thirties only a few large sailing ships remained.

Windmills for mechanical power on land may have first appeared in Persia, where archaeologists have found evidence of the use of wind-driven water pumps for irrigation dating from about the fifth century. These early Persian designs used cloth sails and had a vertical axis, the vertical sails on one side caught the wind while it spilled out on the other side. With the vertical axis there is no difficulty in steering the sails or blades to face the wind. The traditional horizontal axis tower mill for grinding corn, with sails supported by a large tower rather than a single post, had been developed by the beginning of the fourteenth century in several parts of Europe. Its use continued to expand until the middle of the nineteenth century when the spread of the steam engine as an alternative, cheaper, source of power led to its decline. Rural areas in the United States experienced a similar situation at the beginning of the twentieth century. Thousands of farms had steel-towered windmills to pump water and in some cases generate electricity at that time, but over the next fifty years the rural electrification scheme became established and the great majority of these early windmill systems were allowed to fall into disuse. The success of these devices is illustrated by the estimated figures of 50,000 wind/electric generators, or aerogenerators, supplied over the period (1,2).

Historical Development of Wind-Generated Electricity

Denmark

Towards the end of the last century, the windmill was the principal source of power in agricultural areas of Denmark. Known as horse-mills, they were often mounted on the roofs of barns and, together with industrial mills, were estimated to be producing about 200 MW (3) from over 30,000 units. In about 1890, Professor P. La Cour commenced work on wind-power and obtained substantial support from the Danish Government, which not only enabled him to erect a windmill at Ashov, but provided a fully instrumented wind tunnel and laboratory. Between 1890 and his death in 1908, Professor La Cour developed a more efficient, faster running windwheel, incorporating a simplified means

151

of speed control, and pioneered the generation of electricity. The Ashov
windmill had four blades 22.85 m (75 ft) in diameter, on a steel tower
24.38 m (80 ft) high. Power was transmitted, through a bevel gearing, to a
vertical shaft which extended to a further set of bevels at ground level, and
the drive was connected to two 9 kW d.c. generators - the first recorded
instance of wind-generated electricity. By 1910 several hundred windmills of
up to 25 kW capacity were supplying villages with electricity.

The use of wind-generated electricity continued to increase and during the
1939-44 war period, a peak of 481,785 kWh was obtained from 88 windmills in
January 1944 (4). Included among these 88 was the F.L. Smidth unit erected
in Gedser in 1942. Originally a 70 kW d.c. unit, with three 24.38 m (80 ft)
diameter wooden blades, it was converted to a.c. in 1955. It produced
approximately 700,000 kWh during its first five years in operation, or about
2,000 kWh per annum/rated kW.

The United States
By 1922 the Farm Light and Power Year Book listed 54 different manufacturers
of windmill pumps and electrical plant. Towards the end of that decade one
of the most successful windmill manufacturing companies was established - the
Jacobs Wind Electric Company of Minneapolis, Minnesota (5). The Company was
founded by M.L. Jacobs who introduced two significant innovations in his
designs; (a) a three-bladed propellor which effectively eliminated vibrations
found in two-bladed systems due to variations in the total forces acting on
the blades as they moved from the horizontal to the vertical position, and
(b) a flyball governor to control the pitch of the blades, allowing them to
"feather" when the wind velocity was greater than 8.05 m/s (18 mph) and maintain a constant speed to drive the generator. The blades had a diameter of
about 4.27 m (14 ft) and were directly coupled to the generator without
gearing. Perhaps the best-known application was in the Antarctic, where
Admiral Byrd installed a Jacobs generator during one of his scientific
expeditions in the nineteen-thirties. The unit was still functioning when
Byrd returned in 1946. The rural electrification scheme forced the Company
out of business in 1957.

World's Largest Ever Windmill (6)
Conceived by an American engineer, Palmer C. Putnam, in the nineteen-thirties,
the Smith-Putnam windmill had two blades with a diameter of 53.34 m (175 ft)
and was erected at Grandpa's Knob in central Vermont in 1941. The synchronous
electric generator and rotor blades were mounted on a 33.54 m (110 ft) tower
and electricity was fed directly into the Central Vermont Public Service
Corporation network. The windmill was rated at 1.25 MW and worked well for
about 18 months until a main bearing failed in the generator, a failure
unconnected with the basic windmill design. It proved impossible to replace
the bearing for over two years because of the war and during this period the
blades were fixed in position and exposed to the full force of the wind.
During the original assembly of the mainly stainless steel blades and
supporting spars, rivet holes had been drilled and punched in the blades and
cracks had been noticed in the metal around the punched holes in 1942. It
was decided to carry out repairs on site, rather than returning the whole
assembly to the factory. On March 26th 1945, less than a month after the
bearing had been replaced, the cracks widened suddenly and a spar failed
causing one of the blades to fly off. The S. Morgan Smith Company, who had
undertaken the project, decided that they could not justify any further
expenditure on it, apart from a feasibility study on the installation of

other units in Vermont. This indicated that the capital cost per installed
kilowatt would be some 60% greater than conventional systems.

Although sceptics have tended to regard this experiment as an expensive
failure, it was the most significant advance in the history of windpower.
For the first time synchronous generation of electricity had taken place and
been delivered to a transmission grid. Both mechanical failures were due to
a lack of knowledge of the mechanical properties of the materials at that
time. Bearing design and the problems of fatigue in metals have been studied
extensively since then and similar failures are unlikely to occur in modern
windmills. Their research programme included an extensive series of on-site
measurements, which proved that the actual site at Grandpa's Knob had a mean
wind velocity of only 70% of the original estimated velocity* and that many
other sites should have been selected. The technical problems of converting
wind energy into electricity had been largely overcome and the possibility of
developing wind power as a national energy resource in any country with an
appropriate wind climate had been established.

Russia
In 1931, the Russians built the first windmill to feed electricity directly
into an a.c. network at Yalta, near the Black Sea (7). Used as a supplementary power source, it was connected to a conventional fossil-fuel plant at
Sevastopol, about 30 km away. It had three blades 30.48 m (100 ft) in
diameter driving a 100 kW induction generator through wooden gears. The
tower was 30.48 m (100 ft) high but was provided with an inclined strut to
carry the thrust of the wind from the top of the tower to the ground. The
base of the strut was driven round a circular track by an electric motor controlled by a wind direction sensing vane at the top of the tower. The metal-
covered blades could be feathered by an automatic pitch control system
activated by the effect of centrifugal force on offset flaps, so that the
plant could continue to operate in high winds at an approximately constant
speed. An annual output of 279,000 kWh was reported from the site which had
an annual mean wind velocity of 6.7 m/s (15 mph), but satisfactory control
was difficult to achieve. Over the next two decades developments in Russia
were limited to plants generating up to 3 kW (5).

The United Kingdom
By the nineteen-twenties, interest in small wind powered electrical
generators had been well established. Comparative test results on seven
different commercially available windmills ranging in power from 250 W to
10 kW had been published (8) and also a handbook for practical engineers who
wished to build their own machines (9). During the nineteen-thirties, the
Lucas 'Freelite' was developed (10), intended for use with up to six lighting
points - three 40 W and three 25 W bulbs at 25 volts. On the Freelite, the
rotor could be turned out of high winds by means of a furling handle at the
base of the tower.

Two 100 kW machines were build shortly after the war. The first, in 1950,
was built by John Brown and Company and erected in the Orkneys (11), had 3
15.24 m (50 ft) diameter blades mounted on a 23.77 m (78 ft) tower driving a
100 kW a.c. induction generator. The second was built for Enfield Cables by
deHavilland Propellors and the Redheugh Iron and Steel Company and featured a

---

*The measured annual output of about 30% of the predicted performance con-
fixmed the cube law relationship as $0.7^3 = 0.343$.

pneumatic transmission system intented by Andreau in France (12). The two
24.38 m (80 ft) diameter blades were hollow with a hole at the tip so that
during rotation they acted as a centrifugal pump. The induced internal air
flow entered the base of the tower through a turbine, directly coupled to a
synchronous generator. Originally erected at St. Albans in 1953, it could
not be tested there because of the poor wind climate, and was subsequently
re-erected in Algeria for Electricité et Gaz d'Algérie in 1957. Its full
rated output of 100 kW was obtained shortly after re-erection. Studies were
also carried out on the evaluation of windmill performance and a graphical
method for predicting the performance of wind powered electrical generators
was established by the Electrical Research Association (13) in 1960, based on
results obtained from two windmills, a three bladed 12.19 m (40 ft) diameter
unit rated at 25 kW erected on the Isle of Man and a three bladed 10 m
diameter unit rated at 7.5 kW installed in Scotland. Excellent agreement
between the experimental results and the predicted performance was obtained
in each case.

Wind Energy Potential

Wind has a dependable annual statistical energy distribution but a complete analysis of how much energy is available from the wind in any
particular location is rather complicated. It depends, for example, on the
shape of the local landscape, the height of the windmill above ground level
and the climatic cyclé. Somewhat surprisingly, the British Isles have been
studied more extensively than practically any other country in the world (10,
14) and the west coast of Ireland, together with some of the western islands
of Scotland, have the best wind conditions with mean average wind speeds
approaching 9 m/s.

The kinetic energy of a moving air stream per unit mass is $\frac{1}{2}V^2$ and the
mass flow rate through a given cross-sectional area A is ρAV. The theoretical power available in the air stream is the product of these two terms:-

$$\frac{1}{2}\rho AV^3$$

If the area A is circular, typically traced by rotor blades of diameter D,
then $\frac{\pi}{4} D = A$, and the power available becomes:-

$$\frac{\pi}{8}\rho D^2V^3$$

The actual power available can be conveniently expressed (14) as:-

$$K_r D^2V^3$$

where K_r is a term specifically associated with wind dynamics and the
efficiency of the rotor power system.

The maximum amount of energy which could be extracted from a moving air-
stream was first shown by the German engineer Betz, in 1927, to be 0.59259 of
the theoretical available power. This efficiency can only be approached by
careful blade design, with blade-tip speeds a factor of 6 times the wind
velocity. Any aerogenerator will only operate between a certain minimum wind
velocity, the starting velocity V_s, and its rated velocity V_R. Typically

V_R/V_S lies between 2 and 3. If the pitch of the blades can be altered at velocities greater than V_R, the system should continue to operate at its rated output, the upper limit depending only on the design. In some systems the whole rotor is turned out of the wind to avoid damage at high wind speeds. An annual velocity and power duration curve for a continuously generating windmill is shown in Fig. 8.1. Many current designs give a rotor conversion efficiency of 75%.

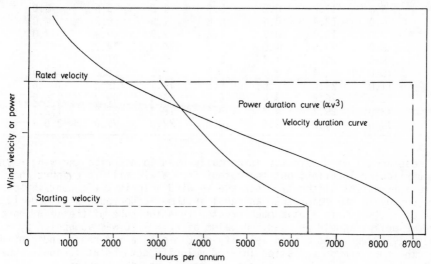

Fig. 8.1.   Annual Velocity and Power Duration
            Curves

Taking the air density, ρ, as 1.201 Kg/m³ at normal atmospheric pressure (1000 millibars) and 290 K and assuming a rotor conversion efficiency of 75%,

$$K_r = \frac{\pi}{8} \times \frac{1.201 \times 0.593 \times 0.75}{1000} \quad \text{(or approximately 0.00020).}$$

The effect of the height of the windmill tower on the performance can be significant and empirical power law indices have been established (15), relating the mean wind velocity V to the height H, in the equation $V = H^a$. A value of a = 0.17 is the accepted value in the U.K. for open, level ground, but this rises to 0.25 for an urban site and 0.33 for a city site. An ideal site is a long, gently sloping hill. Methods for determining the probable mean wind velocity and energy multiplication factors have been described by Caton (16) and Rayment (17).

The following table gives the rotor shaft power available for different wind velocities and rotor diameters, based on the equation $0.0002\ D^2 V^3$ and taking into account a correction factor for the tower height. The actual mechanical or electrical output would be less than this, according to the energy conversion efficiency of each particular system.

The energy produced per annum by a windmill is given by:-

$$E_a = K_r D^2 V^3 \times K_s H \quad kWh$$

where H is the average number of hours in a year (8766) and K_s is a semi-empirical factor associated with the statistical nature of wind energy recovery.

TABLE 8.1

WIND VELOCITY		ROTOR DIAMETER (metres)				
m/s	mph	3.65	5.0	7.0	12.0	18.8
4.0	8.9	0.1	0.3	0.5	1.8	5.0
5.0	11.2	0.2	0.5	1.0	3.6	9.8
6.0	13.4	0.4	0.8	1.8	6.2	17.0
7.0	15.7	0.7	1.3	2.9	9.9	27.0
8.0	17.9	1.0	2.0	4.3	14.7	40.3
9.0	20.1	1.4	2.9	6.1	21.0	57.3
10.0	22.4	2.0	3.9	8.3	28.8	78.6
11.0	24.6	2.6	5.2	11.1	38.3	104.6
12.0	26.8	3.4	6.8	14.4	49.7	135.9
13.0	29.1	4.3	8.6	18.3	63.2	172.7
14.0	31.3	5.4	10.8	22.8	79.0	215.7

The mean annual wind velocity is normally used to describe the wind regime at any particular location, but the output from a windmill is proportional to V^3. Since a transient arithmetic increase in wind velocity will contribute much more energy to the rotor than an equal arithmetic decrease will deduct, the mean of V^3, which is always much greater than the cube of the mean annual wind velocity, should be used. A value of K_S = 1.20 was suggested by Juul (3) in 1956, who used a mean velocity of 8 m/s as a reference and considered that the most common variation in wind velocity occured at frequent short intervals between 6 m/s and 10 m/s, 8^3 equalling 512, whereas $\frac{1}{2}(6^3 + 10^3)$ = 608. A more recent computer analysis by Pontin (18) in 1975 suggests that a value of 2.06 could be taken, giving an approximate value for $K_r K_S$ as 0.0004 and $K_r K_S H$ becomes 3.5064. This value is very close to the figure derived by Rayment, based on data published in Met 0792 (19) and Caton's analysis (16), where the annual extractable energy, if the rotor shaft is connected to an electrical generator, is given by:-

$$E_a = 0.0148 \ V_{50}^3 \ GJ/m^2$$

$$or \ 3.2289 \ D^2 V_{50}^3 \ kWh$$

V_{50} is the velocity exceeded for 50% of the year and is quite close to the mean annual wind speed. Taking 3.5064 and considering the 18.3 metre diameter rotor considered in Table 8.1, the total power produced per annum would be as follows:-

TABLE 8.2

Mean Wind Velocity (m/s)	kWh/annum	Mean Wind Velocity (m/s)	kWh/annum
4	75153	9	856034
5	146782	10	1174258
6	253640	11	1562938
7	402771	12	2027118
8	601220		

Some Recent Developments

The Vertical Axis Windmill

The modern vertical axis windmill is a synthesis of two earlier inventions;
(a) the Darrieus (20) windmill with blades of symmetrical aerofoil cross-
section bowed outward at their mid-point to form a catenary curve and
attached at each end to a (vertical) rotational axis perpendicular to the
wind direction and (b) the Savonius (21,22) rotor or S-rotor, in which the
two arcs of the "S" are separated and overlap, allowing air to flow through
the passage. Simple Savonius rotors have been made out of two standard oil-
drums cut in half and welded together to form the blades (23). A fibre-glass
unit is shown in Fig. 8.2. The Darrieus windmill is the primary power-
producing device, but, like other fixed-pitch high-performance systems, is
not self-starting. The blades rotate as a result of the high lift from the
aerofoil sections, the S-rotor being used primarily to start the action of
the Darrieus blades. The wind-energy conversion efficiency of the Darrieus
rotor is approximately the same as any good horizontal system (24) but its
potential advantages are claimed to be lower fabrication costs and
functional simplicity (25). A major investigation of the system is being
carried out by the Sandia Laboratories (1).

The ERDA Model Zero 100 kW Wind Generator (2,26,27,28)

This is the major project in the United States wind-energy programme and consists of a two-bladed, 38.10 m (125 ft) diameter, variable-pitch propeller
system driving a synchronous alternator through a gearbox, mounted on a
30.48 m (100 ft) steel tower. The test programme has been designed to
establish a data base concerning the fabrication, performance, operating and
economic characteristics of propeller-type wind turbine systems for providing
electrical power into an existing power grid. The blades are located down-
stream from the tower and a powered gear control system replaces the traditional tail fin of earlier designs. Power generation commences when the wind
speed reaches 3.58 m/s (8 mph) and reaches its rated 100 kW at 8.05 m/s
(18 mph), a V_R/V_S ratio of 2.25. The maximum blade rotational speed is
40 rev/min and is maintained at higher wind speeds by varying the blade pitch
angle. This is achieved by a complex hydraulically operated pitch-control
system.

Denmark

The largest windmill system in the world under construction in 1976 (29) was
at Twind, near the west coast of Jutland. The reinforced concrete tower was
completed by December 1975 with a designed rotor axle height above ground
level of 54 m. With two rotor blades, each 27 m long, the starting velocity
is 3 m/s with a rated velocity of 14 m/s. At wind velocities between 14 and
20 m/s, the blade pitch is varied until, above 20 m/s, the system shuts down
with the blades locked and inoperable. Perhaps the most interesting feature
of this project is that it is not sponsored by the Danish government and is
the result of a team effort from a college community who are providing their
own labour and finance, estimated at about $350,000.

The Wind Energy Supply Company, U.K.

A new approach to windmill design has been adopted by the Wind Energy Supply
Company whose system features a simple and robust propeller type variable-
speed automatic windmill rotor which, without the need for any additional
control gear, is self-starting and protected against over-speeding. This
makes it completely safe in gale force gusts and also vibration free. Unlike

Fig. 8.2.

earlier windmills, there is no electrical generation at the mast head.
Instead, there is direct oil/hydraulic high pressure power generated by the
windmill rotor with direct fluid transmission to the point of use. A 5 m
diameter unit is shown in Fig. 8.3. The modular system design may use
commercially available components and means that the system is completely
flexible and can be tailored to individual needs by the choice and arrangement of standard modules. The range of applications for this system is
potentially very large, covering the direct feeding of electricity into a
national grid, through agricultural and horticultural uses, district heating,
desalination, domestic heating and domestic lighting. The windmills could be
provided with automatic storage and standby facilities (a wind/diesel
installation could have an "efficiency" approaching 300% or more from the
point of view of the oil). A prototype 18.3 metre diameter module is due to

Fig. 8.3.

be working towards the end of 1976, initially to provide heat for a large
greenhouse in the south of England and design studies have also been carried
out on a 46 metre diameter module.

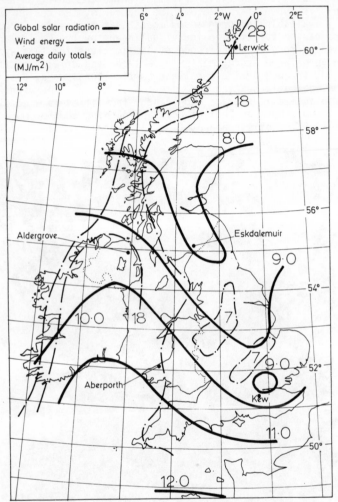

Annual mean global daily irradiation (MJ/m²) and daily wind energy
available at the rotor (MJ/m²)

Fig. 8.4.

Large Scale Wind Energy Programmes

As well as the United States and Denmark, countries such as Sweden, Holland,
Canada, Israel, West Germany and Japan all have plans for wind-driven
electricity generators that could be linked to the national grid. In March,
1976, the Electrical Research Association in England, in evidence to the
House of Commons Select Committee on Energy, stated that giant windmills
could generate up to 10% of the U.K. electricity requirement within ten years
and quoted approximately 1500 windmills as a guideline. The essential
feature of all these various proposals is that they are based on existing
technology, or a technology that is clearly within grasp.

Fig. 8.5.

The Wind-Solar Approach

The complementary nature of wind and solar energy in the British Isles is illustrated in Fig. 8.4, based on approximate maps of solar radiation (30) and windpower (17). This shows that although the annual mean values of global radiation are lower in the north, there is a very much greater potential in wind energy. In many countries it has been found that high winds occur most frequently in the winter months when the demand for energy is also at its peak. It is also the period when solar radiation is at its lowest for direct water or space-heating applications. One solution could be to combine the availability of wind with a solar heating system through the use of ducted rotors. Ducted rotors which are free to rotate into the wind can theoretically extract much more power from a given wind than an unducted rotor of the same diameter. For example, a figure of 46% more power for a 3.5 m diameter ducted rotor has been given by Lewis (31) and 65% by Lilley and Rainbird (32). If these ducted rotors are not placed on a tower, but arranged in fixed banks to form a "wind-wall", an aesthetically satisfying arrangement is obtained which avoids the visual obtrusion of isolated large windmill units into an urban landscape. One such arrangement has been proposed (33) for a housing scheme in Sussex, where long-term heat storage would be provided by underground water storage tanks and the cut for this storage used to provide slopes or banks on which solar collectors can be placed. The wind-wall is placed at the top of the slope, as shown in Fig. 8.5, with a group of garages erected over the underground storage system. The fixed direction of the ducted rotor, together with their height, give an estimated efficiency factor of 77% compared with a similar sized normal windmill.

References

(1) Blackwell, B.F. and Feltz, L.V., Wind energy - a revitalized pursuit, SAND-75-0166, Sandia Laboratories Energy Report (March 1975).

(2) Reed, J.W., Maydew, R.C. and Blackwell, B.F., Wind energy potential in New Mexico, SAND-74-0077, Sandia Laboratories Energy Report (July 1974).

(3) Juul, J., Wind Machines, Wind and Solar Energy Conference, New Delhi, UNESCO (1956).

(4) Golding, E.W. and Stodhart, A.H., The use of wind power in Denmark, ERA Technical Report C/T 112 (1954).

(5) Clark, W., Energy for survival, p. 539, Anchor Press, New York, 1975.

(6) Putnam, P.C., Power from the wind, D. Van Nostrand, New York, 1948.

(7) Gimpel, G., The windmill today, ERA Technical Report IB/T22 (1958).

(8) Cameron Brown, C.A., Windmills for the generation of electricity, Institute for Research in Agricultural Engineering, Oxford University, 1933.

(9) Powell, F.E., Windmills and wind motors, Percival Marshall & Co., London, 1928.

(10) Golding, E.W., The generation of electricity by wind power, E. & F.
      Spon, 1955. Reprinted CTT 1976.

(11) Venters, J., The Orkney windmill and wind power in Scotland, The
      Engineer, 27th January 1950.

(12) Wind generated electricity prototype 100-kW plant, Engineering 179,
      4652, 28th March 1955.

(13) Tagg, J.R., Wind driven generators: The difference between the
      estimated output and actual energy obtained, ERA Technical Report
      C/T 123 (1960).

(14) Golding, E.W. and Stodhart, A.H., The potentialities of windpower for
      electricity generation, British Electrical and Allied Industries
      Research Association, Tech. Rep. W/T16 (1949).

(15) Davenport, A.G., Proceedings of the (1963) Conference on Wind Effects
      on Building and Structure Vol. 1, HMSO (1965).

(16) Caton, P.G., Standardised maps of hourly mean wind speed over the
      United Kingdom and some implications regarding wind speed profiles,
      Fourth International Conference on Wind Effects on Building and
      Structures, London, 1975.

(17) Rayment, R., Wind energy in the U.K., The Building Services Engineer,
      44, 63-69 (June 1976).

(18) Pontin, G.W-W., The bland economics of windpower, Wind Energy Supply
      Company, Redhill (1975).

(19) Tables of surface wind speed and direction over the United Kingdom,
      Meteorological Office, Met 0792, HMSO (1968).

(20) Darrieus, G.J.M., Turbine having its rotating shaft transverse to the
      flow of the current, US Patent 1,835,018, 8th December 1931.

(21) Klemin, A., The Savonius wing rotor, Mechanical Engineering 47, No. 11,
      (November 1925).

(22) Savonius, S.J., The S-rotor and its application, Mechanical Engineering
      53, No. 5 (May 1931).

(23) Brace Research Institute, McGill University, Montreal, Canada.

(24) South, P. and Rangi, R.S., A wind-tunnel investigation of a 14 ft
      diameter vertical-axis windmill, National Research Council of
      Canada, LTR-LA-105 (September 1972).

(25) South, P. and Rangi, R.S., The performance and economics of the
      vertical-axis wind turbine developed at the National Research
      Council, Ottawa, Canada, Agricultural Engineer (February 1974).

(26) Kocivar, B., World's biggest windmill turns on for large-scale wind-
      power, Popular Science (March 1976).

(27) Puthoff, R.L. and Sirocky, P.J., Preliminary design of a 100 kW wind turbine generator, NASA, NASA TM X-71585 (August 1974).

(28) Hamilton, R., Can we harness the wind?, National Geographic (December 1975).

(29) Hinrichsen, D. and Cawood, P., Fresh breeze for Denmark's windmills, New Scientist, 567-570, 10th June 1976.

(30) Solar Energy: a UK assessment, UK Section ISES, London (May 1976).

(31) Lewis, R.I., Wind power for domestic energy, Appropriate Technology for the U.K., University of Newcastle-upon-Tyne (March 1976).

(32) Lilley, G.M. and Rainbird, W.J., A preliminary report on the design and performance of directed windmills, ERA Technical Report C/T 119, (1957).

(33) Environmental Design Group, Wind-Solar Folio, Proposals for the Lewes District Council, Sussex.(1975)

CHAPTER 9

SOME PRACTICAL HEATING APPLICATIONS

Introduction

The previous chapters have shown that it is possible to use solar energy to provide a proportion of the total heating demand in many parts of the world. In high latitudes, however, it is important to appreciate that for many days in the winter months the intensity of the solar radiation is too low to provide any useful heat. There are two main types of solar installation which can be tackled by a competent handyman who has had some experience with basic carpentry and, preferably, some knowledge of standard water pipe fittings. The simplest type of solar installation to construct is for low temperature rise applications, such as in swimming pool heating. Here the requirement is for large areas of simple, unglazed, uninsulated collector. The second type of installation is more ambitious, as it involves the domestic water heating system. The collector panels, which would typically have an area of between 4 m² and 6 m² must be glazed and insulated and there are several other factors to be considered, such as the relative positioning of the various components and the length of the pipe runs.

Swimming Pool and Other Low Temperature Applications

An 'Enclosed' Collector

Although in high latitude countries such as the U.K. the simple, low temperature rise enclosed collector would normally be used for the summer months only, it operates at a high efficiency during this period and is very cost-effective. The capital costs of these systems, excluding labour charges, would be recovered in less than three years when compared with the anticipated savings from most conventional sources. One enclosed collector design which has been tested for over eight years is shown in Fig. 9.1. There are no glass or transparent covers needed, as the temperature rise across the heater is kept as low as possible. Provided the panels are located in a fairly sheltered position they will perform at least as well as a glazed panel, because there is always a radiation loss of approximately 10% in passing through any transparent cover. There is no insulation provided at the sides or the back, again because the temperature in the panel is normally close to ambient temperature and heat losses will be negligible. They are called 'enclosed' as the heated water flows underneath the heat absorbing material and does not evaporate.

Structure

The main structural member of the panel is the backing sheet which can be based on any appropriate flat surface such as plywood sheet, preferably waterproofed. The standard 8 ft x 4 ft plywood sheet has proved easy to work with and provides an area of just over 3 m². The most important feature common to all these low temperature collectors is a thin matt black heat absorbing surface which can absorb nearly all the incident solar radiation. Butyl has proved to be a very satisfactory material for this application and the Butyl used on the original low temperature panels developed by the author

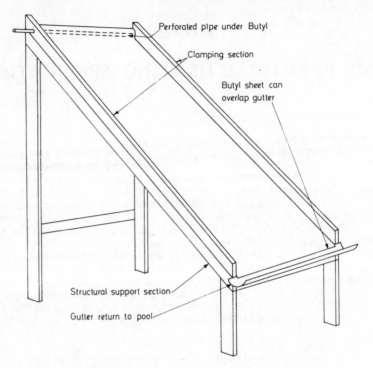

Fig. 9.1 Basic Low Cost Collector

in 1968 (1,2) showed no signs of degradation in 1976. This black surface
sheet is placed on top of a second "water flow spreading" surface so that the
water which is to be heated can flow, under gravity, in a thin layer between
the two sheets.

There are various ways of producing a thin uniformly spreading film of
water on a sloping surface, but one method which has proved successful is to
make the second surface out of a commercial polythene packing material known
as Airwrap. This consists of a uniform matrix of equally spaced cylindrical
air bubbles enclosed in polythene. The major disadvantage of this material
is that its resistance to UV degradation is poor and it has a very limited
life if exposed directly to solar radiation. However, when protected by the
Buryl it has also survived for over eight years. The water inlet at the top
of the collector consists of a small bore perforated pipe. The minimum pipe
diameter should be 15.0 mm and the holes should be at least 2 mm diameter and
spaced about 10 to 15 mm apart. None of these dimensions are very critical
and it is easy to test the pipe before final assembly to check that a uniform
flow is obtained. A straightforward series connection of several panels may
not be completely satisfactory, as the pressures and rate of flow in the
system could mean that progressively less water reached successive panels.
This can be overcome either by a branching system, taking the incoming water
to each end of each panel or by increasing the flow area in the panels with
insufficient water by drilling more holes and/or enlarging the holes. The
heated water is returned to the pool by gravity, so that the bottom of the
panels must be higher than the pool surface. Plastic rainwater gutters make
excellent return channels and the evaporation loss is negligible. Again, it

is easy to test that the slope from the bottom of the panels to the pool is adequate for the necessary flow.

List of Materials

(1) Flat panel for backing - ⅜ inch weatherbonded ply is satisfactory (length L, width W).

(2) Butyl sheet.

(3) Airwrap packing sheet.

(4) Inlet pipe - plastic, 15 mm diameter is adequate. Length as needed for connecting to adjacent panels.

(5) Plastic rainwater gutter for return flow to pool. Length as needed for connecting to adjacent panels and return flow to pool.

(6) Structural supporting sections, two of length L, three or four of width W. Cross-section not critical, but adequate for rigidity.

(7) Top clamping sections of length L, primarily for sealing the edges.

(8) Framework to support panel, as required.

(9) A stop valve and appropriate flexible piping to connect with panel inlet pipes.

Lengths L and W could be a nominal 8 ft x 4 ft.

Brief Construction Details
The airwrap polythene is stretched over the plywood with the bubbles facing upwards and held in place at the turned-over ends by a few drawing pins. The Butyl is laid over it and clamped by the section (7) above to the structural support, as shown in Fig. 9.2. The main structural members are fitted, then the perforated pipe. It is useful to test the pipe at this point, before fixing the top end of the Butyl sheet to overlap it, to ensure that an even flow of water is obtained. The return flow gutters should be fitted last when the panels are sited, as it is necessary to have them sloping gently towards the pool.

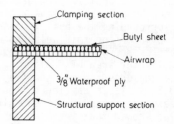

Fig. 9.2. Detail of Clamping on Sides

Flow Rate
The temperature rise must be kept as low as possible to reduce heat losses. One litre of water heated from $15^{\circ}C$ to $55^{\circ}C$ has received only one tenth the amount of heat supplied to 400 litres of water heated from $15^{\circ}C$ to $16^{\circ}C$, although the former would be immediately detected and pronounced hot! Flow rates should therefore be at least 150 litres/hour/m^2 of collector or about 3 U.K. gallons/hour/ft^2 of collector. It is important to ensure that the collector panel is not distorted and that the entry pipe is horizontal. Failure to check these points could lead to the water 'streaming' to one side or the other, greatly reducing the overall effectiveness of the system as heat from the solar radiation can only be transferred efficiently to the

water when it is directly in contact with the Butyl sheet. A separate pump
is often not needed as the collector system can be connected by a T-pipe to
the return pipe from an existing filtration system.

Position and Direction

The angle of tilt and the direction which the panel faces are not very
critical. South facing panels are considered ideal for the Northern hemisphere, but a few degrees either side of south will make very little
difference. For the summer heating season only, a fairly well sloped panel
is best, perhaps 40^0 to the horizontal or less. It is possible to provide
complete computer simulations for the various periods of the year to predict
the optimum angle of tilt, but local conditions often impose greater limitations, e.g. the presence of large screening trees or buildings. The use of
the roof of an existing building to mount the collectors will often be very
convenient and can reduce the visual impact of a large expanse of black
surface. Common sense should indicate any unsuitable, partially shaded location, but it is advisable to check for over-shading on any site over a full
day in the early part of the heating season. The author has seen at least
two installations which were shaded from direct sunlight for the greater part
of the afternoon.

Size, Performance and Economics

The ratio of the collector area to the pool surface area is a convenient
ratio to start with. This ratio has been used (3,4) to estimate the likely
daily temperature rise in a pool under various radiation conditions. To
obtain a temperature rise of about 5^0C in one day under good summer conditions in a temperate climate a collector/pool surface area approaching
1.5:1 has been suggested, but it is important to appreciate that a steady
increase in the pool temperature over a period of several weeks early in the
season can be achieved with a ratio as low as 0.25:1. This is because the
ground surrounding the pool is also heated by the relatively higher temperature pool water and this helps to maintain the pool at comfortable swimming
temperatures during periods of several consecutive cloudy days. Even with a
solar collector area of one tenth the pool surface area, sufficient energy
could be collected in one good day to give an additional temperature rise of
about 0.5^0C. Tests carried out during the 1975 swimming season at a school
in Sussex, where panels based on the author's design principles were
installed showed that very substantial savings had been achieved compared
with the previous season. In 1974, with electric pool heating only, the
swimming season lasted from the end of May until the first week of September
and 48,885 units of electricity (kWh) were used (5). In 1975 with the solar
panels fitted to supplement the electric heating, the swimming season lasted
from mid-May until October, but the electricity consumption was reduced to
14,232 units. Average pool temperatures between 23^0C and 29^0C were obtained.
These figures would need to be checked against the mean total radiation
values for the periods concerned, but is is reasonable to assume that about
500 kWh/m^2 were collected in 1975, the balance coming from radiation directly
onto the pool surface.

The cost of materials for the Butyl, airwrap and wooden-framed solar water
heater is less than £10 (at 1976 values) and this gives a payback period of
about two years on the basis of savings in heating costs compared with conventional methods.

Other Designs - The 'Open' System

With a convenient south-facing (in the northern hemisphere) corrugated roof,

such as galvanised iron, a perforated pipe can be laid along the ridge and
the poolwater can be pumped up to the pipe and allowed to flow down the
corrugations. In these systems the flowing water is open to the atmosphere
and some heat losses due to evaporation are inevitable, making their overall
efficiency perhaps only two-thirds that of the enclosed type. The holes in
the perforated pipe should be placed opposite each groove in the corrugated
sheet with a minimum diameter of about 5 mm, as the distance between holes
will be at least 75 mm. Similar flow rates to the enclosed system should be
used to keep the temperature rise low. The efficiency of this system can be
improved by stretching a clear plastic material such as the DuPont "Tedlar"
PVF (polyvinyl fluoride) film type 400 BG20 TR, over the corrugations,
turning it into a type of Thomason system. Alternatively, a standard clear
outdoor corrugated plastic can be used.

    Another 'open' system which has been tried successfully in Sussex (6) con-
sists of a large flat black stepped concrete area. The pool water is pumped
to the top step and cascades gently over the black concrete to the pool.
This is a very simple system, easy to construct and relatively large
collector/pool surface area ratios can be achieved. The only difficult
feature in the construction is to obtain a uniform thin film of water over
the entire surface area. The use of a long flexible perforated pipe at the
top step helps to achieve this.

Controls
Sensitive differential-temperature on-off control systems are not really
necessary for these low temperature applications. It has been found to be
perfectly adequate to control the system manually, allowing the water to flow
through the panels from about 08.00 to 18.00 hours every day except when it
is very cloudy and overcast. If a differential-temperature controller is
used, it should have some type of time-delay in the circuit, so that continuous cycling does not occur in intermittent cloud conditions.

Pool Cover
Before constructing a swimming pool solar heater remember that it is much
easier and far more cost-effective to provide a pool cover to reduce the
greatest source of heat loss from the pool - evaporation. The simplest
method is to use some type of floating cover. Light-gauge black polythene
sheet will make a difference. It can be clamped round the edges of the pool
and needs only small holes at approximately 0.3 m intervals to allow surface
rain water to pass through. Commercial floating pool covers are often made
of two layers of blue PVC, separated by strips of polyurethane foam. Overnight temperature drops are usually in the order of 1°C compared with over
2°C for the uncovered pool. This difference of 1°C sounds very small, but
represents over 100 kWh in a small, 20,000 (UK) gallon pool.

Domestic Solar Water Heaters

    There are probably hundreds of different types of solar water heater
design, all of which make some contribution to the hot water requirements of
their particular installation. Most designs, however, have certain essential
features in common and these are as follows:-

(1) The collector plate.

(2) Insulating material at the back and sides of the plate.

(3) One or two sheets of glass or translucent plastic in front of the
        collector plate.

(4) A casing to contain items 1, 2 and 3.

(5) A hot water storage system, which may be a separate storage tank.

   The major British work in this field was carried out by the late
Professor Harold Heywood, between 1947 and 1955, and his design principles
have subsequently formed the basis for many types of solar heating system
(7,8).

Casing, Covers and Insulation

The purpose of the casing and glass or plastic cover(s) is to act as a
weatherproof container for the collector plate. Any conventional flat box-
like shape can be modified to hold the collector plate and support the
cover(s). Glass reinforced plastic (GRP) is a fairly popular choice for the
casing, but wood or sheet metal can also be used. Examples of typical cross-
sections illustrating each type are shown in Fig. 9.3.

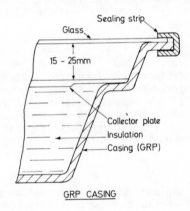

GRP CASING

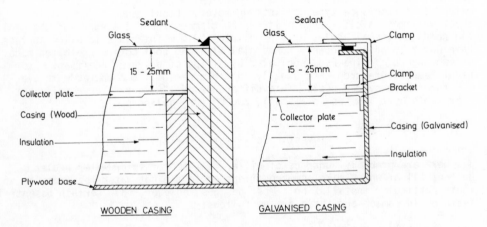

WOODEN CASING   GALVANISED CASING

Fig. 9.3.

A single sheet of glass will transmit about 90% of the incident solar radiation but will trap nearly all the heat radiated by the collector surface, as glass is opaque to long wave radiation. The use of two covers further reduces the amount of radiation which reaches the collector plate, but if the collector plast is more than about 35°C above the temperature of the surroundings, a second cover improves the collector efficiency as it reduces the heat losses from the outer cover to the surroundings. It also helped to protect the plate from overnight freezing temperatures in winter. The increased cost and difficulty of fitting a second cover, as well as the com- paratively small performance advantage obtained in U.K. conditions, make a single cover the recommendation for the simple practical collector system.

As an alternative to glass, the use of translucent plastic can be con- sidered, provided that is has been specially treated to stand up to the weather. An early weather resistant material, DuPont 'Mylar', has been replaced by Tedlar PVF type 400 BG20 TR, and is used in several commercially available collectors in the USA. This is a thin film material and can be heat sealed, shrink wrapped or bonded by adhesives. Glass-reinforced trans- lucent plastic (GRP) sheets have been used successfully both in the UK (9) and the USA (10). GRP is easier to handle than glass, especially if working on an exposed roof as considerable care has to be taken with glass to avoid breakages. The distance between the parallel surfaces of two covers or a single (or inner) cover and the collector plate should be between 15 and 25 mm. The exact spacing is not very critical (11). For cheapness, with comparatively little loss in overall performance, 4 mm horticultural grade glass can be used for the cover. One reservation about the use of plastic materials is that even those which have been specially developed to withstand weathering will have a limited lifetime and the manufacturers should be con- sulted directly about the likely life. In designing the detail of fitting and sealing the cover to the casing, avoid leaving a water trap at the edges. Some designs have ignored this and consequently have a small pool of dirty water almost permanently on the lower edge of the collector cover surface. Various commercial insulating materials can be used at the rear and sides of the collector plate, providing they can stand up to maximum temperatures of over 100°C - quite possible with the system not operating and exposed on a hot sunny day. Fibreglass or mineral (rock) wool is very satisfactory but polystyrene should not be used as an insulating material, as it melts if placed in close contact with a hot collector plate. A minimum thickness of 50 mm of insulation is suggested at the rear of the collector plate, with a minimum of 25 mm at the sides although this is not so critical and could be omitted.

Collector Plate Designs

General

Selective surface. The provision of a selective surface is beyond the scope of the great majority of practical home workshops. Copper is probably the easiest material to coat, using warm solution of sodium hydroxide and sodium chlorite, the temperature and concentration of which should be care- fully controlled (12). Even with the commerically available coatings, there are conflicting views about how long they continue to be effective. It can be seen from the performance curves in Chapter 3 that selective surfaces have an advantage only when the temperature of the collector is relatively high.

Frost protection. The problem of freezing in the winter can be dealt with in several ways. The simplest is to forget about solar heating for the entire winter period and drain the collectors. The amount of heat not collected between mid-October and mid-March would represent perhaps 20% of the total annual potential energy gain. If an anti-freeze solution is used, the system must be completely self-contained and should have the approval of the Local Water Authority. Such systems use an indirect hot water tank fitted with a heat transfer coil connected directly to the solar collectors and are discussed later.

Corrosion. The problems of corrosion were also dealt with in Chapter 3. One cause of trouble is likely to be a combination of mixed metals in the system, such as copper and aluminium, either directly in contact under moist conditions, or if an aluminium collector plate with hollow passages is subjected to ordinary mains water containing certain dissolved chemicals. This was highlighted at the ISES Congress at Los Angeles in 1975 where there was an almost unanimous view from the exhibitors of solar panels (13) that the problems of possible corrosion and leakage in systems containing aluminium collectors were so serious that aluminium could no longer be considered as an appropriate collector material. Although there may be no direct contact between dissimilar metals in a solar water heating system, corrosion problems may occur whenever copper and steel or galvanised components are used, particularly when the water is cupro-solvent. In the early stages of use, copper from the collector plate or connecting pipes could dissolve in the water and be redeposited in the galvanised storage tank. Similarly a galvanised steel collector plate could corrode if it were connected to a copper tank.

The presence of dissolved oxygen in the system is another equally important factor in corrosion, but can be completely overcome by the use of an all-copper system. Copper, which is very widely used in general plumbing practice, does not corrode in oxygenated water or in a suitably treated anti-freeze solution.

Pipe runs. For normal installations operating on a thermosyphon system, a pipe diameter of 28 mm is recommended between the collector and the storage tank. Normal good plumbing practice should be followed in all pipe runs, keeping right angle bends to a minimum, especially in thermo-syphon systems. The main problem likely to arise is the formation of air locks in the system. When ordinary mains water is heated, dissolved air comes out of solution and the slow accumulation of bubbles at any point establishes an air lock which either completely stops the flow or reduces the circulation rate. It is essential that there should be a continuous rise in both the flow and return pipes from the collector to the storage tank. It is not essential but can be advantageous for the collector to be slightly inclined so that the horizontal header sections rise gently towards the collector outlet. Adequate provision should be made for vent pipes in the system. Pipe lengths between the various component parts in the system should be as short as possible. All heated pipes should be insulated.

Specific

Corrugated galvanised collector. Heywood's first practical domestic collector, installed in his home near London in 1948, consisted of two sheets of corrugated galvanised steel placed in 'mirror image' position so that eight water channels were formed along the length of the collector surface.

The edges were rivetted and soldered, and had square section headers fitted at the top and bottom edges. The front surface of the collector, which had an area of just under 1 m², was painted matt black and enclosed in a wooden frame with two sheets of glass over the front and insulating material at the back. He commented that while the collector worked will in a conventional thermo-syphon system for a number of years 'it did not have a long life' (7).

Nevertheless, a modified version of the original Heywood collector has been developed very successfully by the Brace Research Institute (14). The Brace Collector was designed to incorporate low-cost materials generally available, even in relatively remote parts of the world, and is based on two galvanised steel 22 gauge sheets, one being corrugated and forming the heat absorbing surface and the other being flat. The two sheets are rivetted and soldered together, the corrugated surface is painted black and the collector is placed in a simple galvanised steel box lined with an insulating material - coconut fibre is suggested. A single sheet of 3 mm window glass is fitted using a silicone sealant and allowing a 3 mm clearance all round for glass expansion. Hot water storage is in a converted 45 gallon oil drum. An initial life of about five years with negligible maintenance is claimed and more than seven years satisfactory operation was reported for some units in Barbados.

Pipes bonded to a metal sheet. The Commonwealth Scientific and Industrial Research Organisation (CSIRO), Australia, published a guide to the principles of the design, construction and installation of solar water heaters in 1964 (12) which was summarised in the JIHVE in 1967 (15). The absorber plate described in their guide consisted of pipes thermally bonded to a metal sheet. Copper was their preferred material and a framework of 28 mm and 15 mm copper pipes were soldered to a 26 gauge copper sheet (approximately 0.45 mm), the vertically rising 15 mm pipes being brazed to the 28 mm header pipes. Galvanised steel or asbestos cement was recommended for the collector casing at that time. Slightly thicker copper sheet, of 20 to 24 gauge (approximately 0.56 to 0.91 mm) is currently recommended in the UK. A plan view of a typical pipe matrix is shown in Fig. 9.4. This could be connected or bonded to a flat or corrugated sheet. The recommended distance between the centre line of adjacent pipes is about 150 mm.

Although copper is the preferred material, galvanised or aluminium pipes or sheet could be used. Any departure from the excellent thermal bond which can be achieved by the copper-copper system will be less efficient, the worst bond being achieved by simple wire tying at large intervals. The Henry Mathew collector (16) however, shown in Fig. 9.5, achieves good results with a theoretically poor collector, as the distance between wire ties is about 750 mm. In this all steel system, the galvanised pipe in the collector is horizontal, but adjacent pipes are much closer together than the recommended 150 mm.

The use of clips at shorter intervals will improve the bond, and forming a semi-circular depression in a flat sheet to fit the pipe will also add to the performance, particularly if some form of adhesive or filler is than added, if soldering is not appropriate.

As an alternative to the matrix system, a continuous serpentine loop can be used, as shown in Fig. 9.6. This idea is used in some commercial panels, but its use is limited to forced circulation systems.

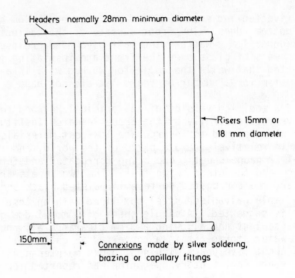

Fig. 9.4. Pipe Matrix

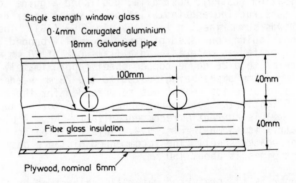

Fig. 9.5. Section of Henry Mathew Collector

Commercial steel panel radiators. The standard commercial steel panel
radiator can be easily modified to make a solar collector plate. If possible,
the panel should be obtained without its final coat of glossy white paint, as
the absorbing surface should be painted with a standard matt black paint.
The panel radiator first tested by the author in 1968 still had its original
matt black painted surface intact after eight year's exposure under a single
Mylar transparent cover sheet (17). The rear side of the panel can be left
white. There are normally four connexions at the top and bottom of the panel
at each end. The cold return water inlet should be arranged to enter a
bottom connexion and the solar heated water should leave from the diagonally
opposite connexion, i.e. bottom left to top right or bottom right to top
left. Do not connect the inlet pipe by a branching connexion to each end of
the panel or try to take the heated water from both top outlets, as this can
reduce the overall efficiency. In the thermosyphon system, for example, an
internal circular flow pattern could be established with a double inlet and

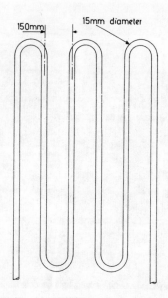

Fig. 9.6. Serpentine System

outlet arrangement. The painted radiator, with its connexions ready, should
be placed in the casing with its ribs following the normal pattern, i.e.
running upwards to the horizontal header as shown in Fig. 9.7. The panel
will work if turned at right angles, but with an overall loss of efficiency.

A simple trough collector/storage tank. An effective collector can be
made out of a watertight casing with a sloping bottom as shown in Fig. 9.8.
This is a combined collector and solar storage tank and is particularly suitable where there is only an existing cold water supply. It will not be suitable for poor radiation conditions or freezing ambient temperatures. The
cold inlet water displaces the heated water at the shallower end of the
casing when the inlet control valve is opened. The use of the sloping bottom
ensures that after a brief period of good radiation conditions there is a
layer of heated water available. A few small ventilation holes should be
drilled in the casing below the glass cover to minimize the effect of
moisture condensation. GRP painted matt black or lined with Butyl is
suggested for the casing material. A simpler version consisting of a square-
sectioned uniform depth galvanised casing was described and tested by the
National Building Research Institute, Pretoria in 1967 (18).

Size, Performance, Economics and Storage Capacity

Heywood's measurements (19) showed that in the UK at a latitude of $51^{\circ}31'$
North, a south facing surface inclined at an angle of 40° to the horizontal
would receive a daily average of 9.2 MJ/m^2 or about 2.56 kWh/m^2. (See
Chapter 2 for a detailed analysis for other angles and periods of the year.)
Differences of a few degrees in the angle of inclination or in direction from
south make very little difference and this figure can be used as the basis
for an estimate of the average amount of useful heat which can be supplied to
a domestic hot water system. Laboratory tests can give values of collector

Fig. 9.7.

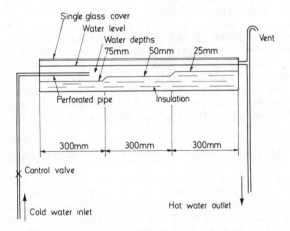

Fig. 9.8. Flat Trough Collector/Storage

efficiencies well over 60% for moderate temperature differences, but when due
allowance is made for the longer pipe runs in an actual installation, the
intermittent nature of the sunshine during the day and the way in which water
is used, an overall figure of between 30% and 40% is realistic for the UK.
This means that an annual total of between 280 and 376 kWh can be obtained
per square metre of collector. The figure of 280 kWh was confirmed in a
series of tests carried out for the year September 1973 - August 1974 (20),
although this was regarded as a low figure for two reasons - it was a below
average year for solar radiation and the storage tank was not well insulated.
The Building Research Establishment have quite independently suggested (21,
22) figures of 324 kWh/m^2 for a 6 m^2 installation and 350 kWh/m^2 for a 4 m^2
installation. There is an important feature to appreciate about these overall values for efficiencies between 30% and 40%. Any increase in collector
area above 6 m^2 for an average household will not give a proportionate
increase in the total amount of useful heat collected. If it did, then an
area of about 12 m^2 would provide hot water all through the year. This is
impossible because of the very low levels of radiation received in the winter
period. Heywood's average daily figure is 1.05 kWh/m^2 between October 16th
and February 26th, and for most of December and January this figure would be,
at best, only half this value. So to approach the averaged daily demand of
about 10 kWh towards the mid-Winter period, a collector area of about 50 m^2
would be necessary, but even this would probably not work, as there is
another practical barrier - all flat plate collectors have some lower limit
of radiation level below which they cannot operate. The best figure to take,
therefore, lies between the limits of 280 and 375 kWh/m^2, although particular
installations in certain parts of the country might do considerably better
than this. Annual savings for 4, 5 and 6 m^2 of collector, taking the figure
of 324 kWh/m^2, would be as follows:-

TABLE 9.1

Annual Savings (£/annum)

Unit cost of one kWh (pence)	Collector area (m^2)		
	4	5	6
2	25.92	32.40	38.88
3	38.88	48.60	58.32
4	51.84	64.80	77.76

Material costs of practical systems (at 1976 prices) excluding labour,
should be under £30/m^2, giving a payback period of about five years based on
a replaced energy charge of 0.02 £/kWh.

An exception to this general rule of between 4 m^2 and 6 m^2 of collector
area was seen in the Copper Development Association's 1975 prize winning
entry, where 8 m^2 was used (23). The designer subsequently commented that he
felt it was quite wrong to have an underpowered system (24).

The ratio of approximately 50 litres of hot water storage capacity for
every square metre of collector area (one UK gallon/ft^2), originally suggested
by Heywood, has been accepted as the standard for the great majority of
domestic solar heating systems.

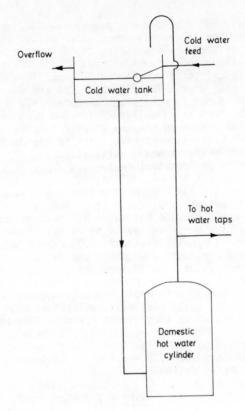

Fig. 9.9. A Standard Domestic System

Integrating Solar Water Heating into the Domestic Hot Water System

The Thermosyphon System

The basic components in a standard domestic system are shown in Fig 9.9. The simplest method of using solar water heaters is in the direct natural thermo- syphon system, shown in Fig. 9.10, using a separate solar hot water storage tank. As the water is heated in the collector it rises to the top of the collector and then passes to the upper section of the storage tank. As the same time the cooler water at the bottom of the storage tank returns to the bottom of the collectors. As the flow is caused by the difference in density between the hot and cold water there must be a difference in level, H, between the bottom of the storage tank and the top of the collector. With a minimum of 600 mm it is unlikely that any reverse flow at night could take place, but a non-return valve could be fitted in the cold water flow pipe. Chinnery (25) reported that any reduction of H below 600 mm led to a corres- ponding reduction in the overall efficiency of the system as indicated in Table 9.2.

The hot water pipe from the collectors should enter the upper section of the storage tank at a level lying between two-thirds and three-quarters of the total capacity. Again, Chinnery shows that loss of efficiency occurs when this entry point is lowered. The use of a separate storage tank has an

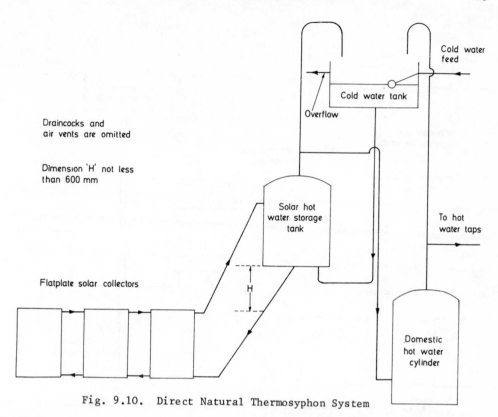

Cold water
feed

Cold water tank

Overflow

Draincocks and
air vents are omitted

Dimension 'H' not less
than 600 mm

Solar hot
water storage
tank

To hot
water taps

Flatplate solar collectors

H

Domestic
hot water
cylinder

Fig. 9.10. Direct Natural Thermosyphon System

advantage in countries such as the UK because for the greater part of the year
the temperatures that can be achieved from the solar collectors are too low
for direct use in the domestic hot water system. It is useful to arrange for

TABLE 9.2

	Mean Efficiency
Collector outlet at 600 mm below bottom of tank	54.6%
Collector outlet level with bottom of tank	46.4%
Collector outlet about $\frac{2}{3}$ tank height above bottom	43.8%

the possibility of by-passing the hot water cylinder during good radiation
conditions in the summer so that conventional heating can be turned off.
This is particularly useful if no hot water is drawn off during the day, as
otherwise, in the evening, there could be a mixing of very hot solar heated
water with the unheated water in the hot water cylinder. With the separate
solar hot water storage tank, even a small temperature rise above the cold

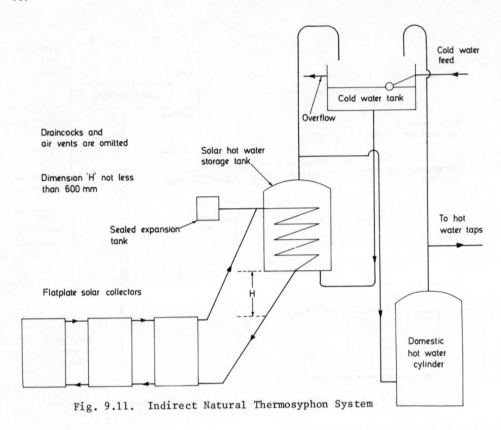

Fig. 9.11. Indirect Natural Thermosyphon System

water tank temperature can save energy, as this heated water enters the hot
water cylinder instead of the direct cold feed water. The thermosyphon
system can also be used indirectly, as shown in Fig. 9.11. A heat exchanger
is fitted inside the solar hot water storage tank and there is a closed
circuit from the collectors through the heat exchanger. This circuit contains a sealed expansion tank, or provision for a separate cold water
"topping-up" tank and an overflow pipe. The sealed expansion tank prevents
fresh oxygen from getting into the system and inhibits corrosion. The
indirect circuit can be filled with an anti-freeze solution, but this must be
completely self-contained so that the anti-freeze does not leak into the
domestic system. A heat exchanger can be a simple coil of copper pipe and
some commercial groups recommend a standard small domestic copper cylinder as
the storage tank/heat exchanger unit. These are not designed for operating
at the lower temperature differences and flow rates likely to be encountered
in a solar installation and it is far better to use about one metre of finned
28 mm diameter copper pipe for 200 litres of storage tank. With indirect
circuits care should be taken to connect the hot water pipe from the
collectors to the top of the heat exchanger. The system could work with the
hot water pipe connected to the lower end of the heat exchanger, but it would
be considerably less effective, as the water flow rates would be much lower.

Forced Circulation or Pumped Systems
As can be seen from Fig. 9.12, a forced circulation system is more complicated and the circulating pump, which can be a normal small central heat-

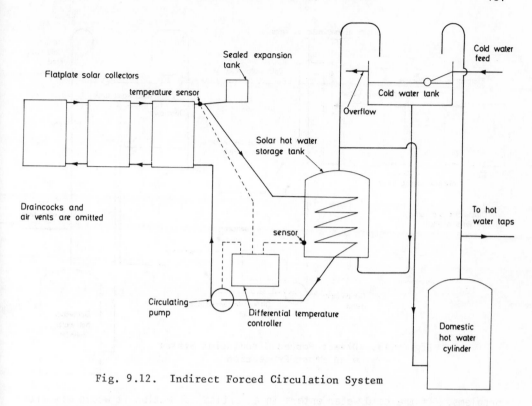

Fig. 9.12. Indirect Forced Circulation System

ing pump, must be controlled by a differential-temperature controller.
These are supplied by various solar collector manufacturers or can be built
from an electronic circuit design (25,26). An analysis of the operation of a
differential-temperature controller and the type of problem encountered has
been given by O'Connel (27), who warns that on days with comparatively low
radiation intensity and intermittent cloud cover, a system could cycle continuously, loosing more energy than is gained. The setting of the temperature difference for switching on and off the pump is important, as is the
position of the temperature sensors, which should not be placed at a high
level on the storage tank.

An alternative method of protecting the system against frost damage is
shown in Fig. 9.13, where the collectors can be completely drained by a
solenoid-operated drain valve. This system is more elaborate than the others
and care must be taken in deciding the height and position of the various
components, so that the cold water tank does not empty when the drain valve
is opened.

A system which eliminates the separate solar hot water storage tank is
shown in Fig. 9.14. The combined domestic/solar cylinder has two heat
exchanger coils in it, the upper one being connected directly to a conventional boiler system. The advantages are that less space is taken up and
fewer pipes are used, but under ideal conditions the efficiency will be less
than with a separate storage tank.

The entry of the cold water supply to the hot water cylinder can create

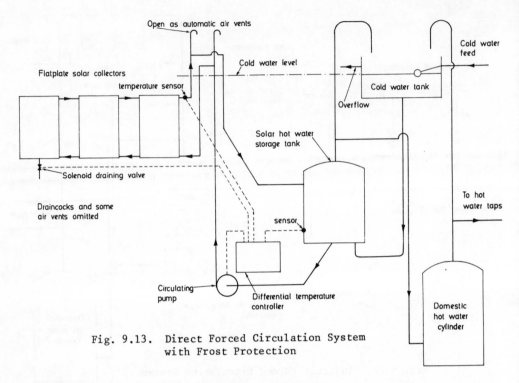

Fig. 9.13.   Direct Forced Circulation System
             with Frost Protection

problems. If the cold water enters in a vertical direction it would mix with
the heated water near the top of the cylinder, cooling it very rapidly. One
way of overcoming this is to position the entry pipe so that it discharges
horizontally or slightly downwards into the tank. The cold water supply
should never be connected to the return pipe of a direct thermosyphon system.
In principle this could be satisfactory during the day, as the cold water
would absorb heat as it passed through the collector plate, but at night the
cold water would also enter the hot water cylinder near the top and mix
directly with heated water.

Local Building and Planning Regulations

Solar heating systems must comply with local building and planning
regulations. For example, if a collector is erected on a roof or fixed to a
house it must be secure and not liable to be blown off in a high wind. It is
also probable that some planning authorities would raise objections if solar
heating panels made a substantial alteration to the visual appearance of a
building. This would be particularly relevant in the case of older buildings
of historic interest. Many people might object to the rather stark appearance of swimming pool heaters and it may be necessary to site these behind a
hedge or similar screening in conditions which are somewhat less than ideal.

Checking the Claims of Commercial Brochures

As yet there are no national standards for solar heating systems in the
UK and many misleading statements have been published. Typical examples are

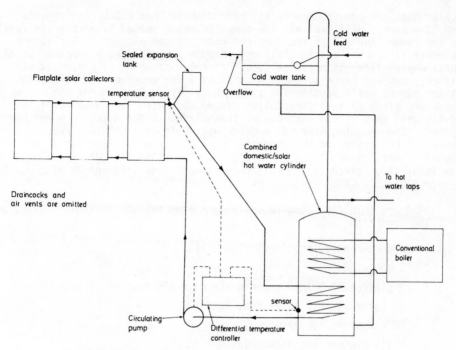

Fig. 9.14.   Single Cylinder Indirect Forced
Circulation System

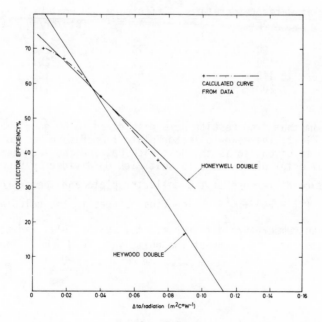

Fig. 9.15.

"solar heat can provide nearly all your domestic hot water requirements free"
and "the panels will heat all the domestic water needed by an average family
in the summer and 80% of the required water in the winter". This is not
necessarily untrue, but it could only happen if people were prepared to alter
their way of life quite radically and face the prospect of storing dirty
dishes, cups and saucers, dirty clothes etc. for weeks on end during the
winter months while waiting for a few sunny days. Even in the summer months
there are often several consecutive cloudy days with low solar radiation
levels when very little heat can be transferred to the domestic water heating
system. Some manufacturers do provide useful independent test results and
these can be plotted on the characteristic performance chart for single and
double covers, shown in Fig. 9.15. The chain-dotted line was calculated from
the information given in the following extract from a recent UK brochure by
Senior Platecoil Ltd:-

NASA Lewis in Cleveland recently completed testing the Solar Collector.
The conditions were:

1. 300 Btu/hr/sq. ft. input flux level

2. 7 mph wind

3. 2 glasses of $\frac{1}{8}$ inch green glass with 88% transmission level

4. 10 lb/hr/sq. ft. water flow rate

5. 80°F ambient (temperature)

6. Plain brushed on black paint

The results, considered to be good, were as follows:

Inlet water temperature °F	Efficiency %	Heat Pick Up (Btu/hr/sq. ft.)
80	70	210
100	67	201
140	56	168
200	38	114

Can their claim that "the results were considered to be good" be sub-
stantiated? First, the mean collector plate temperature has to be calculated.
The flow rate is given, so is the Heat Pick Up, and the mean temperature rise
across the collector plate, Δt_c, is Heat Pick Up divided by Flow Rate. The
mean temperature difference of the collector plate and the surroundings, Δt_a,
is given by $(T_i - 80) + \frac{\Delta t_c}{2}$. The final stages in the calculation are to
divide the mean temperature difference, Δt_a, by the total radiation landing
on the collector and then convert the units into $°C \ W^{-1} \ m^2$. The new table is
then as follows:-

TABLE 9.3

Inlet Water (Ti) (deg F)	Efficiency %	Δt_c (deg F)	Δt_a (deg F)	$\dfrac{\Delta t_a}{300}$	(^{O}C W^{-1} m^2)
80	70	21	10.5	0.035	0.0061
100	67	20.1	30.05	0.010	0.0176
140	56	16.8	68.4	0.228	0.0401
240	38	11.4	125.7	0.419	0.0738

The points joined by the chain-dotted line on Fig. 9.15 lie very close to the Honeywell double glazed collector performance and are better then Heywood's early work, so the manufacturer's claim that the results were considered to be good is quite justified. Very few collectors have been independently tested in this way. The claims by a few manufacturers to have achieved annual savings in the UK approaching 1000 kWh per m² of collector area in a domestic hot water system have never been verified.

Other Energy Saving Methods

Although the provision of adequate loft insulation and trying to eliminate draughts by sealing round the edges of windows and doors will not be as interesting or exciting as building a solar water or space heating system, these simple measures will be far more cost effective at present. An analysis carried out in the UK in 1974 (28) gave the following figures for capital cost and estimates fuel savings over a five year period:-

	Cost of Installation	Estimated value of saved fuel in 5 years
Basic Roof Insulation (50 mm)	£30	£110
Draught Prevention	£10	£50

This should be compared with solar heating at 1976 prices

	Cost of Installation	Estimated value of saved fuel in 5 years (with no inflation)
Practical 6 m² system (excluding labour)	£180	£200
Commercially available 6 m² system	About £500	£200

References

(1) Dewhurst, J. and McVeigh, J.C., A low-cost solar heater, The Heating
     and Ventilating Engineer, March 1968.

(2) McVeigh, J.C., Some experiments in heating swimming pools by solar
     energy, J.I.H.V.E. 39, pp 53-55, June 1971.

(3) How to heat your swimming pool using solar energy, Brace Research
     Institute, McGill University, January 1965, revised February 1973.

(4) deWinter, F. and Lyman, W.S., Home built solar water heaters for
     swimming pools, ISES Congress 'The Sun in the Service of Mankind'
     Paris, 1973.

(5) Plumb, M., Solar tanning for swim pool heating bill, Sussex Express and
     County Herald, 28th May 1976.

(6) Carter, The Hon. Mrs. B., Brencar Solar Exports Ltd., Rogate, Hants.

(7) Heywood, H., An appraisal of the use of solar energy, Society of
     Engineers 57, p 155 (1966).

(8) Heywood, H., Solar Energy: Past, present and future applications,
     Engineering 176, p 409 (1953).

(9) Brachi, P., Sun on the roof, New Scientist, 19th September 1974.

(10) Scoville, A.E., An alternative cover material for solar collectors,
     ISES Congress, Los Angeles, Extended Abstracts, Paper 30/11,
     July 1975.

(11) Duffie, J.A. and Beckman, W.A., Solar Energy Thermal Processes, John
     Wiley & Sons, New York, 1974.

(12) Solar Water Heaters, CSIRO Division of Mechanical Engineering,
     Circular No. 2, 1964.

(13) McVeigh, J.C., Advances in Solar Energy, Heating and Ventilating News,
     September 1975.

(14) How to build a solar water heater, Brace Research Institute, McGill
     University, February 1965, revised February 1973.

(15) Solar Water Heaters, a summary of (12), J.I.H.V.E. p 309, January 1967.

(16) Reynolds J.S. et al. The Atypical Mathew Solar House at Coos Bay,
     Oregon, ISES Congress, Los Angeles, Extended Abstracts, Paper 43/12,
     July 1975.

(17) McVeigh, J.C., Developments in solar energy utilisation in the United
     Kingdom, ISES Congress, Los Angeles, Extended Abstracts, Paper 10/4,
     July 1975.

(18) Richards, S.J. and Chinnery, D.N.W., A solar water heater for low-cost
     housing, National Building Research Institute, Bulletin 41, CSIR
     Research Report 237, Pretoria, South Africa 1967.

(19) Heywood, H., Operating experiences with solar water heating, J.I.H.V.E.
     39, pp 63-69, June 1971.

(20) Harris, J., The British solar panel is born, Building Services Engineer
     42, p 432, October 1974.

(21) Building Research Establishment, Energy Conservation: a study of energy
     conservation in buildings and possible means of saving energy in
     housing, BRE Current paper CP 56/75, 1975.

(22) Courtney, R.G., An appraisal of solar water heating in the UK, BRE
     Current paper CP 7/76, 1976.

(23) Awards for Solar Heating, Architects Journal, 10th September 1975.

(24) Swinburne, A., CDA Solar Heating Forum, Southampton 1976, reported in
     Heating and Ventilating News 19 (7)(1976).

(25) Chinnery, D.N.W., Solar Water Heating in South Africa, National
     Building Research Institute, Bulletin 44, CSIR Research Report 248,
     Pretoria, South Africa, 1967.

(26) Brad, Differential Temperature Controller Plan, Eithin-y-Caer,
     Churchstoke, Powys.

(27) O'Connel, J.C., The problems associated with the use of differential
     temperature controllers, Solar Energy for Buildings Seminar, North
     East London Polytechnic, February 1976.

(28) Booker, C. and Gray, B., Inflation? But we are burning money, The
     Observer, London, 6th October 1974.

APPENDIX 1

SOME USEFUL UNITS, DEFINITIONS AND CONVERSIONS

Systeme International (SI) units have been adopted by many countries including the United Kingdom. Some papers on solar energy, particularly the earlier references, use other systems of units. The following definitions and conversion factors may be used:-

Length

1 millimetre (mm)	0.0393701 inch (in)
1 metre (m)	3.28084 ft
1 Angstrom (Å)	10^{-10} m

Area

1 square centimetre (cm^2)	0.155000 in^2
1 square metre (m^2)	10.7639 ft^2

Volume

1 cubic centimetre (cm^3)	0.0610237 in^3
1 litre	10^3 cm^3 or 10^{-3} m^3
1 Imperial gallon (UK)	4.54596 litre
1 US gallon	3.78531 litre

Mass

1 kilogramme (kg)	2.20462 lb
1 tonne (10^3 kg)	0.984207 ton (UK)

Pressure

1 kg/cm^2	14.2233 lb/in^2

Heat, Energy and Power

The British Thermal Unit (Btu) is the amount of heat required to raise 1 lb of water through 1 degree Fahrenheit.

The Calorie is the amount of heat required to raise 1 gm of water through 1 degree Celsius (Centigrade).

The Langley is a unit of energy frequently used in radiation work and is equivalent to 1 calorie/cm^2.

Heat is a form of energy and the Joule (J) is commonly used as a mechanical unit of heat. The fundamental unit of power is the Watt (W).

1 Btu = 1.05506 x 10^3 joule (J) = 778.169 ft lb

1 calorie (cal) = 4.1868 J

1 kilocalorie = 3.96830 Btu

1 Watt = 1 Joule per second = 0.00134 horsepower

1 kilowatt hour (kWh) = 3.600 x 10^6 J = 3.600 MJ = 3.41213 x 10^3 Btu

1 Btu/h = 0.293071 W

1 kilocalorie/m^2 = 0.368668 Btu/ft^2 = 1.163 W/m^2

1 W/m^2 = 3.6 kJ/m^2/h = 0.316998 Btu/ft^2/h

1 Btu/h ft^2 $^{\circ}$F = 5.67826 W/m^2 $^{\circ}$C

Some Thermal Energy Equivalents of Fuels

Oil   1 barrel = 0.63 x 10^{10} J, 1 ton = 4.4 x 10^{10} J

Black Coal 1 ton   = 2.9 x 10^{10} J

Natural Gas 1 ft^3 at STP = 1.05 x 10^6 J

Preferred Abbreviations in SI Units

Giga	G	10^9	milli	m	10^{-3}
Mega	M	10^6	micro	μ	10^{-6}
Kilo	k	10^3	nano	n	10^{-9}

UK OBSERVING NETWORK AND THE STORAGE OF DATA

The primary solar radiation network in the UK is the responsibility of the Meteorological Office. Monthly mean values of global radiation, diffuse radiation and illumination on a horizontal plane appear fairly soon after observation in the Monthly Weather Summary. These data are provisional data, and are subject to slight revision after appropriate calibration and consistency checks. The Radiation Section at the Meteorological Office processes the observed data, and when they are satisfied, hourly data are transferred onto magnetic data tapes containing radiation, illumination and sunshine data only, based on local apparent time. Daily totals are summarized on published data sheets which appear about two to three years after actual observation.

Enquiries for more detailed information about the UK radiation network and the associated data banks should be addressed to The Meteorological Office, Eastern Road, Bracknell, Berkshire, RG12 2UR, Telephone Bracknell 20242.

A list of observing stations reporting hourly totals of solar radiation as supplied by the Meteorological Office, is set out in the following table with the following abbreviations:-

T.   Total (global) solar radiation on a horizontal surface.

D.   Diffuse solar radiation on a horizontal surface. (Global radiation with the direct component from the sun removed by a shadering.)

L.   Total illumination on a horizontal surface. (Measured by an illuminometer with a spectral response similar to a human eye.)

F.   Diffuse illumination on a horizontal surface.

B.   Radiation balance. (Incoming minus outgoing radiation of all wavelengths.)

N.S.E.W. Global solar radiation on vertical surfaces facing North, South, East and West respectively.

SS.   Duration of bright sunshine in hours, measured by a Campbell-Stokes sunshine recorder.

Values of T., D., B. and N.S.E.W. are expressed in milliwatt-hours per square centimetre, while L. and F. are expressed in kilolux-hours.

The time standard used throughout is Local Apparent Time (L.A.T.).

<u>UK STATIONS REPORTING HOURLY TOTALS OF SOLAR RADIATION</u>
(EXCLUDING OCEAN WEATHER SHIPS)

Station	Latitude	Longitude	Elevation above sea level (metres)	Elements measured	Date of first observation
Lerwick	60°08' N	01°11' W	82	T.D.L. B.SS. (L. - 1.1.58, B. - 1.1.64)	Jan 1952
Aberdeen	57°10' N	02°05' W	35	T.	June 1967
Dunstaffnage	56°28' N	05°26' W	3	T.	April 1970
Dundee (Mylnefield)	56°27' N	03°04' W	30	T. B.	July 1973
Eskdalemuir	55°19' N	03°12' W	242	T.D.L. B.SS. (L. - May 58, B. - Feb 64)	Jan 1956*
Aldergrove	54°38' N	06°13' W	71	T.D.L. B.SS. (L. - June 71)	Jan 1969
Cambridge	52°13' N	00°06' W	23	T.D. SS.	Jan 1952 - Dec 1971
Aberporth	52°08' N	04°34' W	115	T.D. SS.	July 1957θ
Cardington	52°06' N	00°25' W	29	T.D. SS.	Jan 1972
Hurley	51°32' N	00°49' W	43	T	Mar 1969
London Weather Centre	51°31' N	00°07' W	77	T.D.L. SS. (L. - 1.1.67)	Jan 1958φ
Kew	51°28' N	00°19' W	5	T.D.L.F.B.SS. (F. - Mar 64)	Jan 1947Δ
Bracknell	51°23' N	00°47' W	73	T.D.L.F. SS. (N.S.E.W. - Jan 67)	Feb 1965
Jersey	49°12' N	02°11' W	85	T.D.L. B.SS. (L. - Jan 69)	Jan 1968

*Data from June 1952 in manuscript
θData from Jan 1953 in manuscript   : nil Mar - Dec 1958
φData from Jan 1950 in manuscript
ΔData from July 1932 in manuscript

APPENDIX 3

SOME REFERENCE SOURCES

National Reports

(1) Solar Energy : a UK assessment
UK Section of ISES, c/o The Royal Institution, 21 Albemarle Street,
London W1X 4BS (May 1976).

(2) Description of the solar energy R & D programs in many nations
F. de Winter and J.W. de Winter eds. Available from National Technical
Information Service, U.S. Department of Commerce, 5285 Port Royal Road,
Springfield, VA 22161 (February 1976).

(3) Solar Energy for Ireland
Report to the National Science Council by Eamon Lalor, Government
Publications Sale Office, G.P.O. Arcade, Dublin 1 (February 1975).

(4) Report of the Committee on Solar Energy Research in Australia
Australian Academy of Science Report No. 17 (September 1973).

(5) Solar Energy as a National Energy Resource
NSF/NASA Solar Energy Panel Report. Available from National Technical
Information Service, U.S. Department of Commerce, 5285 Port Royal Road,
Springfield, VA 22161, Report PB 221659 (December 1972).

Books

(1) Solar Energy
H. Messel and S.T. Butler Eds. Shakespeare Head Press, Sydney (1974)
and Pergamon Press (1976).

(2) Solar Energy Thermal Processes
J.A. Duffie and W.A. Beckmann, John Wiley and Sons (1974).

(3) Solar Energy for Man
B.J. Brinkworth. The Compton Press (1972).

(4) Direct use of the sun's energy
Farrington Daniels. Yale University Press (1964) and reprinted
Ballantine Books, New York (1974).

Journals

(1) Solar Energy
The Journal of Solar Energy Science and Technology, published bimonthly
by Pergamon Press for the International Solar Energy Society.

(2) Sun at Work in Britain
     The magazine of the UK Section of ISES, c/o The Royal Institution,
     21 Albemarle Street, London W1X 4BS.

(3) Gelioteckhnika
     The Russian journal published in English translation by Allerton Press,
     New York.

(4) Bulletin of the Cooperation Mediterraneanne pour l'Energie Solaire
     Published by COMPLES, 32 Cours Pierre-Puget, 13006 Marseilles, France.

General Publications on Energy Resources

(1) How to use Natural Energy
     Conservation Tools and Technology Ltd., 143 Maple Road, Surbiton,
     Surrey KT6 4BH. A very comprehensive guide which includes details of
     equipment and summaries of over 50 publications (June 1976).

(2) The Generation of Electricity by Wind Power
     E.W. Golding, E. & F. Spon (1955) and revised edition, Conservation
     Tools and Technology Ltd. (1976).

(3) Aspects of Energy Conservation
     I.M. Blair, B.D. Jones and A.J. Van Horn Eds. Pergamon Press (1976).

(4) Energy for Survival
     Wilson Clark. Anchor Press/Doubleday, New York (1975).

Conference Proceedings since 1974

(1) Wind Energy Systems
     Cambridge, UK (September 1976). British Hydromechanics Research
     Association, Cranfield, Bedford.

(2) Solar Use Now - A Resource for People
     The extended abstracts of the ISES World Congress in Los Angeles (1975).
     Available from most national ISES sections.

(3) Conference Proceedings of the UK Section of ISES
     Cover many applications including the low temperature thermal collection
     in the UK, photovoltaic cells, processing meteorological data, architec-
     ture and planning, European solar houses, biological conversion systems
     and agriculture. Available from the UK Section of ISES.

Information in the UK

(1) The UK Section of ISES can be contacted c/o The Royal Institution,
     21 Albemarle Street, London W1X 4BS.

(2) A Solar Energy Information Office is being established under Department
     of Industry funding to serve industry with data and advice at the
     Department of Mechanical Engineering and Energy Studies, University
     College, Cardiff CF1 1XL.

(3) The Building Research Establishment have a major solar energy research
     programme at Garston, Watford WD2 7JR and publish Current Papers on
     topics related to energy use in buildings.

(4) Airwrap packing sheet - a UK supplier is Abbotts Packaging Ltd., Gordon
     House, Oakleigh Road South, London N.11. Telephone: 01-368-1266.
     (Smallest bubble, grade C120). 1976 cost approximately 25p/m^2. (May be
     difficult to purchase in small quantities.)

(5) Butyl sheet - a UK supplier is Butyl Products Ltd., 11 Radford Crescent,
     Billericay, Essex. Telephone: Billericay 53281. (30 thousandths of an
     inch thick.) 1976 cost approximately £2.00/m^2.

(6) Copper - advice on properties and on the availability of copper based
     solar heating panels and systems from The Copper Development Association,
     Orchard House, Mutton Lane, Potters Bar, Hertfordshire EN6 3AP.
     Telephone: Potters Bar 50711.

(7) Copper pipe - matrix systems and serpentine shapes
     See "Wednesbury copper tube and components for solar heating panels"
     The Wednesbury Tube Company, Oxford Street, Bilston, West Midlands,
     WV14 7DS. Telephone: Bilston 41133.

GLOSSARY

Absorptivity
The ratio of the amount of radiation absorbed by a body to the radiation incident upon it. (Absorptivity = 1 - Reflectivity)

Air mass
The ratio of the actual distance traversed through the Earth's atmosphere by the direct solar beam to the depth of the Earth's atmosphere, normal to the surface.

Albedo
The ratio of the radiation reflected from the Earth to the total amount of radiation incident upon it.

Algae
Filamentous or unicellular water plants, normally fast growing.

Altitude
The angle between a straight line from the sun to the centre of the earth and the tangent to the surface of the earth (90^0 - zenith).

Anaerobic fermentation
Fermentation caused by micro-organisms (bacteria) in the absence of oxygen.

Aphelion
The point in the orbit of the Earth where it is at its greatest distance from the sun.

Array
A bank or set of solar modules or panels.

Attenuation
The reduction of radiation flux over a given path length, due to absorption and scattering.

Autarchic
Self-sufficient. Applied to a house, it would be independent of all mains services.

Azimuth angle
The angle between the south meridian, measured in a horizontal plane westwards, and the direction of the sun (note that this convention is normally used in solar work).

Bioconversion
The conversion of solar energy into chemically stored energy through biological processes. Various fuels and materials can be produced by bioconversion.

Black body
A term denoting an ideal body which would absorb all and reflect none of the
radiation falling upon it. Its reflectivity would be zero and its absorptivity would be 100%. An alternative definition is a body which at any one
temperature emits the maximum possible amount of radiation, i.e. its
emissivity is 1.0. The total emission of radiant energy from a black body
takes place at a rate expressed by the Stephen-Boltzmann Law.

Celestial Sphere
A sphere of infinite radius with its centre located at a point within the
solar system. The reference frame for all systems of astronomical spherical
coordinates are based on the celestial sphere.

Collector
A solar collector or absorber is used to collect solar radiation. In the
process, the radiation undergoes a change in its energy spectrum.

Concentration ratio
The ratio of the irradiance at the focus of the concentrator to the direct
radiation received at normal incidence on the surface.

Concentrator
A device for focussing solar radiation.

Cuprosolvent
Tends to dissolve copper.

Declination
Solar declination is the angular distance between the sun and the plane of
the celestial equator.

Diffuse radiation or insolation
Solar radiation which arrives on Earth as a result of the scattering of
direct solar radiation by water vapour and other particles in various layers
of the atmosphere. It is also known as indirect radiation or sky radiation.

Direct radiation or insolation
Radiation from the sun which eventually reaches the Earth's surface without
being scattered. This is also known as direct beam radiation.

Emissivity
The ratio of the radiation emitted by a body to that which would be emitted
by an ideal black body the same temperature and under identical conditions.

Evacuated tubular collector
A collector manufactured from three concentric glass tubes. The central tube
contains the heat transfer fluid, while the evacuated space between the outer
two tubes reduces heat losses. Characteristically, their efficiency is in
the order of 50% with temperature differences between the heated fluid and
the surroundings greater than $100^{o}C$.

Flat plate collector
Any non-focussing flat surfaced solar collector.

Global radiation
The sum of the intensities of the vertical component of direct solar radiation and the diffuse radiation. An alternative definition is radiation at the Earth's surface from both sun and sky.

Greenhouse effect
An expression given for the solar heating of bodies shielded by glass or other transparent materials which transmit solar radiation but absorb the greater part of the radiation emitted by the bodies.

Heat pipe
A device for transferring heat by means of the evaporation and condensation of a fluid in a sealed system.

Heat pump
A reversed heat engine; it transfers heat from a lower temperature to a higher temperature by the addition of work. The amount of heat delivered at the higher temperature divided by the work input is called the coefficient of performance.

Heliostat
A mobile array of mirrors used to reflect a beam of sunlight in a fixed direction as the sun moves across the sky.

Incidence angle
Angle between the perpendicular to the surface and the direction of the sun at that instant.

Infra-red radiation
The band of electromagnetic wavelengths lying between the extreme of the visible region (circa 0.76 μm) and the shortest microwaves (circa 1000 μm)

Insolation
Originally defined as one of the processes of weathering, it is now generally regarded as another term for solar radiation, including the ultra-violet and visible infra-red radiation regions. Total insolation is a term sometimes used instead of global radiation.

Irradiance
Radiant energy passing through unit area per unit of time.

Irradiation
The process of exposing to radiation.

Perihelion
The point in the earth's orbit when the earth is closest to the sun.

Photobiology
A biological subject covering the relationship between solar radiation, mainly in visible wavelengths, and biological systems.

Photochemistry
A chemical subject dealing with chemical reactions induced by solar radiation.

Photosynthesis
The conversion of solar energy by various forms of plant and algae into organic material (fixed energy).

Photovoltaic cell
Also known as photocell or solar cell. A semiconductor device which can convert radiation directly into electromotive force. An alternative definition is a device used for detection and/or measurement of radiant energy by the generation of an electrical potential.

Power tower or solar tower
A tall tower, perhaps 500 m high, positioned to collect reflected direct solar radiation from an array of heliostats. The top of the tower contains the heat exchange chamber and the hot working fluid is used in a conventional electrical generating system at ground level.

Pyranometer
An instrument used for measuring global radiation, also known as a solarimeter.

Pyrheliometer
An instrument used to measure the direct irradiance of the sun along a surface perpendicular to the solar beam. Diffuse radiation is excluded from the measurement.

Radiant energy
Energy transmitted as electromagnetic radiation.

Radiation
Radiant energy.

Reflectivity
The ratio of the radiation reflected from a body to the radiation incident upon it. (Reflectivity = 1 - Absorptivity)

Retrofitting
Erecting a solar collecting system on to existing buildings.

Scattering
Interaction of radiation with matter where the direction is changed but the total energy and wavelength remain unaltered. An alternative definition is the attenuation of radiation other than by absorption.

Selective surface
A surface which has a high absorptivity for incident solar radiation but also has a low emissivity in the infra-red region.

Semiconductor
An electronic conductor whose resistivity lies in the range 10^{-2} to 10^9 ohm-cm (between metals and insulators).

Solar constant
The amount of solar radiation which is received immediately external to the Earth's atmosphere and incident upon a surface normal to the radiation taken at the mean Earth-sun distance. It is not a true constant as it varies,

mainly due to sun-spot activity. A mean value of 1.35 kw/m^2 ± 4% is normally taken, although values outside these limits can occasionally occur.

Solar furnace
A device used for achieving very high temperatures by the concentration of direct radiation.

Solarimeter
Another name for Pyranometer.

Solar pond
An artificially enclosed body of water containing a stratified salt solution. Solar energy can be stored as heat in the pond, as the stratified salt solution reduces heat losses.

Spectrometer
An instrument for the measuring of radiation intensity over small wavelength intervals.

Spectroradiometer
An instrument for measuring spectral irradiance.

Thermosyphon
Natural liquid circulation caused by the small difference in density between a hot and a cold liquid. In a solar collector thermosyphon system, the collector is placed below the water storage tank and the solar heated water rises to the top of the tank, displacing colder water from the bottom of the tank to the bottom of the collector.

Turbid atmosphere
An atmosphere containing particulate material (aerosols).

Turbidity
A factor describing the attenuation of direct radiation due to suspended particulate material. Sometimes defined as the factor by which the air mass is multiplied to allow for attenuation in the turbid atmosphere.

Ultra-violet radiation
The band of electromagnetic wavelengths lying next to the visible violet (0.10 μm to 0.38 μm).

Visible region
The range between the ultra-violet and infra-red regions. This region can also be defined as that which affects the optic nerves (0.38 μm to 0.76 μm).

Zenith angle
The angle between a line from the sun to the centre of the Earth and the normal to the surface of the Earth (90° - altitude).

LIST OF SYMBOLS

Most of the symbols used in this book are given below. Other symbols, including various constants, are defined in the text.

A   area

C   capital cost

D   diffuse solar radiation, diameter

E   internal energy

e   emmitance

F   collector heat removal factor, annual savings

F_c cost of competitive energy

f   annual inflation rate

G   global solar radiation

G_c incident solar radiation normal to collector

G_i total annual incident solar radiation normal to collector

H   average hours in the year

I   direct solar radiation

i   net effective interest rate

ℓ loss of efficiency factor

M   capital repayment factor

N   mean length of day

n   mean daily bright sunshine hours, a period of years

Q   useful heat collected per unit area, heat transferred

r   annual interest rate

T   temperature, maintenance cost

U   overall heat loss coefficient

V   velocity

W	external work
Y	constant annual payment
Z	original annual payment
z	zenith distance

Greek

α	absorptance
γ	solar altitude
η	efficiency as defined in the text
ρ	reflectance, density
τ	transmittance

Subscripts

1	initial, source
2	final, sink
a	ambient
i	inlet
m	mean, per m^2
R	rated
S	starting

INDEX